Mathematical Economics

I0890974

Mathematics and mathematical economics are subjects foreign to many students. The concepts covered in the average course are lost on many without the opportunity for interactive practice. Matlab and Mathematica® are software packages that enable students to see mathematics come alive on their screens. *Mathematical Economics* serves as a guide to the essential mathematics of economics and as an introduction to the powerful tools represented by those software packages.

This textbook, designed for a single-semester course, begins with basic set theory, and moves briskly through fundamental, exponential and logarithmic functions. Limits and derivatives finish the preparation for economic applications, which are introduced in chapters on univariate functions, matrix algebra, and the constrained and unconstrained optimization of univariate and multivariate functions. The text finishes with chapters on integrals, the mathematics of finance, complex numbers, and differential and differences equations.

Rich in targeted examples and explanations, *Mathematical Economics* offers the utility of a handbook and the thorough treatment of a text. While the typical economics text is written for two-semester applications, this text is focused on the essentials. Instructors and students are given the concepts in conjunction with specific examples and their solutions. Typical mathematical economics texts do not introduce software packages; *Mathematical Economics* introduces two, Matlab and Mathematica. It is the pairing of specific textbook treatments with interactive software that sets this text apart and allows it to be used in a variety of course formats and as a reference.

Arsen Melkumian is Assistant Professor of Economics at Western Illinois University.

Mathematical Economics

Arsen Melkumian

Routledge
Taylor & Francis Group
LONDON AND NEW YORK

First published 2011
by Routledge
4 Park Square, Milton Park, Abingdon, Oxon OX14 4RN
605 Third Avenue, New York, NY 10017

Routledge is an imprint of the Taylor & Francis Group, an informa business

Typeset in Times New Roman by Sunrise Setting Ltd, Devon, United Kingdom

British Library Cataloguing in Publication Data
A catalogue record for this book is available from the British Library

Library of Congress Cataloging in Publication Data
Melkumian, Arsen, 1969–
Mathematical economics / by Arsen Melkumian.
p. cm.
1. Economics, Mathematical. I. Title.
HB 135.M45 2010
330.01'51–dc22
2010021858

ISBN: 978-0-415-77686-8 (hbk)

ISBN: 978-0-415-77687-5 (pbk)

ISBN: 978-0-203-83721-4 (ebk)

Contents

Preface

This book is intended to cover most of the mathematics that would be typically used in undergraduate and lower-level graduate economics courses. The background recommended for readers would be college-level algebra and the knowledge normally acquired in courses on the principles of economics. This book contains numerous examples, visually stimulating graphs and representations. I want especially to point out the need to integrate mathematical knowledge with functional software in today's ever-evolving economic landscape. At the end of many of the chapters I include examples utilizing Mathematica or Matlab. I chose Mathematica over other software packages because of its ability to combine highly sophisticated functions with wonderful usability.

One of the aims of this book is to give aspiring economists the basic mathematical tools they need to excel at all levels of their discipline. I expect the students to learn economic theory in other courses using the mathematical and programming skills acquired in the process of studying this text. The material starts out with basic set theory and runs the gamut of mathematical economics, including univariate and multivariate optimization and some mathematics of finance. Each section comes with its own carefully chosen examples, which teach the material without overwhelming the reader with useless repetition. The logical progression of topics covered should help the reader to master the material presented in this book.

The answers to odd-numbered problems are included at the end of the book. The instructor's manual, which includes answers to the even-numbered problems, is available for those professors, lecturers, and teachers who choose to adopt this textbook.

Acknowledgements

I would like to thank my colleagues from the Departments of Economics and Mathematics at Western Illinois University for performing various editorial tasks. I am especially grateful to my students, who have provided valuable comments and spotted errors. I would also like to express my deep appreciation to my family and friends for making this text possible.

Finally, I would like to thank the editors at Routledge for their excellent guidance.

1 Introduction

The purpose of this chapter is to familiarize the reader with basic concepts in set theory and introduce our notation. The first section introduces sets of natural, rational, irrational and real numbers and provides several examples of relations. The following section considers the polynomial functions frequently used by economists, linear, quadratic and cubic, and demonstrates how to minimize a quadratic profit function (or any quadratic function for that matter). The concepts introduced in this chapter will be used frequently, so a thorough understanding of the chapter's material is essential to the remainder of this book.

1.1 Basic set theory

The numbers that we use for counting, namely $1, 2, 3, \ldots,$ are called *natural numbers*. Number 1 is the lowest natural number and there are infinitely many natural numbers. A collection of natural numbers, such as $1, 7, 25$ and 46, is referred to as a set.

In general, a *set* is a collection of objects. These objects may be natural numbers, coins, movies, people and so on. The objects in a set are referred to as the elements of the set. In some cases we can define a set by enumeration of the elements. For instance, natural numbers $2, 7, 11$ and 23 can form a set

$$S = \{2, 7, 11, 23\}$$

In this case, we have defined our set S by enumeration of the elements. The set S is finite, as it contains a finite number of elements. When a set is infinite or contains a lot of elements, we are forced to define it by description. For example, the set

$$W = \{x \mid x \text{ is a natural number greater than } 5\}$$

reads as follows: "W is the set of all natural numbers x, such that x is greater than five." The vertical bar in the description of W simply means "such that." Set W is infinite, as there are infinitely many natural numbers greater than five.

If the elements of a set can be counted using natural numbers $1, 2, 3, 4$ and so on, then the set is considered to be countable. In particular, set S as well as any other finite set is countable. Infinite sets may be countable or uncountable. For example, set W is countable since its elements can be counted using natural numbers: the first element of W is 6, the second element of W is 7 and so on. The set of natural numbers, referred to as $\mathbb{N}$ and given by

$$\mathbb{N} = \{x \mid x \text{ is a natural number}\}$$

is also countable.

Now, let us define the set of integers

$$\mathbb{Z} = \{\ldots, -3, -2, -1, 0, 1, 2, 3, \ldots\}$$

Clearly, $\mathbb{Z}$ is an infinite set that is countable. The set of integers $\mathbb{Z}$ is also known as the set of whole numbers. We shall write $\mathbb{Z}^+$ or $\mathbb{Z}^-$ to refer to the set of positive or negative integers, respectively.

EXAMPLE 1.1 Show that the set of all integers $\mathbb{Z}$ is countable.

Solution: The set of integers can be counted as follows: the first element of $\mathbb{Z}$ is 0, the second element of $\mathbb{Z}$ is 1, the third element of $\mathbb{Z}$ is -1, the fourth element of $\mathbb{Z}$ is 2, the fifth element of $\mathbb{Z}$ is -2 and so on. The set is countable.

Now, consider another set:

$$\mathbb{Q} = \left\{ \frac{p}{q} \,\middle|\, p \text{ and } q \text{ are integers with no common divisors} \right\}$$

This is called the set of rational numbers. Rational numbers are simply fractions in lowest terms. It could be shown that the set of rational numbers is countable. The set of rational numbers $\mathbb{Q}$ contains both the set of integers $\mathbb{Z}$ and the set of natural numbers $\mathbb{N}$. Alternatively, we can say that the set of integers $\mathbb{Z}$ is a subset of $\mathbb{Q}$ and the set of natural numbers is a subset of $\mathbb{Q}$ and write

$$\mathbb{Z} \subset \mathbb{Q}$$

and

$$\mathbb{N} \subset \mathbb{Q}$$

to indicate that fact. In general, if every element of set A is also an element of set B, we write

$$A \subset B$$

to indicate that A is a subset of B. Further, set A is said to be a proper subset of set B if

$$A \subset B \text{ and } A \neq B$$

Note that $\mathbb{N}$ and $\mathbb{Z}$ are proper subsets of $\mathbb{Q}$. Also, the set

$$S = \{2, 7, 11, 23\}$$

is a proper subset of $\mathbb{Q}$. In fact, since every element of A is also an element of A, we have $A \subset A$.

In set theory, there is a unique set $\varnothing$ that contains no elements. This set is called the empty set and we can easily prove that the empty set is a subset of any set.

EXAMPLE 1.2 Show that the empty set $\varnothing$ is a subset of an arbitrary set A.

Solution: The empty set contains no elements. Therefore we can say that if a is an element of the empty set, then a is also an element of set A and write

$$\text{if } a \in \varnothing \quad \text{then } a \in A$$

The sign $\in$ reads as "belongs to" and

$$a \in A$$

reads as "a belongs to A" or "a is an element of A" or simply "a is in A." We write

$$a \notin A$$

to indicate that a does not belong to A. Integers and rational numbers can be represented on a number line like the one shown in Figure 1.1.

Since there are lots of rational numbers, it would appear that rational numbers could cover the number line. In other words, it would appear that every point on the real line could be assigned to a rational number. The ancient Greeks found that not every number on the number line is rational. In fact, it could be shown that $\sqrt{3}$ is not a rational number. Furthermore, there are infinitely many numbers on the number line that are not rational.

Numbers on the number line that are not rational are called *irrational*. There are infinitely many irrational numbers and the set of irrational numbers is not countable. Intuitively speaking, the set of irrational numbers contains a lot more elements than the set of rational numbers. The *set of real numbers* $\mathbb{R}$ consists of rational and irrational numbers and is uncountable: there is no way to associate elements of $\mathbb{R}$ with the elements of $\mathbb{N}$. In other words, the set of real numbers $\mathbb{R}$ is the union of the set of rational numbers $\mathbb{Q}$ and the set of irrational numbers. There is also a set of complex numbers, which contains all real numbers. Complex numbers are introduced as pairs of real numbers. Complex numbers are defined in Chapter 11 and most of this text deals with real numbers only. Unless otherwise specified, from now on by a "number" we shall understand a real number.

By definition, the *union of two sets* A and B, denoted $A \cup B$, is the set of elements contained in either A or B. Thus:

$$A \cup B = \{x \mid x \in A \text{ or } x \in B\}$$

For example, if $A = \{2, 3, 7\}$ and $B = \{3, 11, 12\}$ then $A \cup B = \{2, 3, 7, 11, 12\}$.

The *intersection of two sets* A and B, denoted by $A \cap B$, is the set of elements contained in both A and B. Thus:

$$A \cap B = \{x \mid x \in A \text{ and } x \in B\}$$

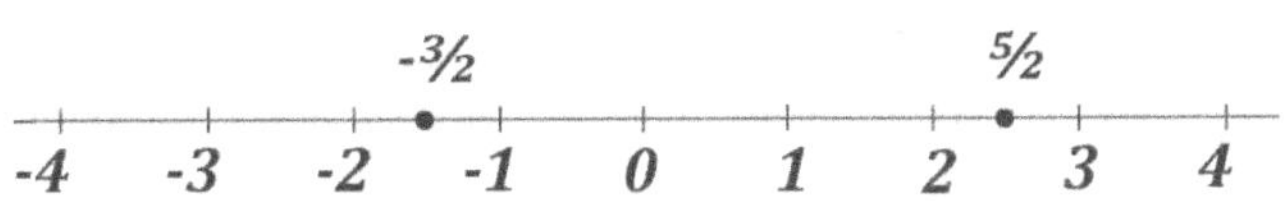

Figure 1.1

For example, if $A = \{2, 3, 7\}$ and $B = \{3, 11, 12\}$, then $A \cap B = \{3\}$. Sets A and B have one element in common. If two sets have no elements in common ($A \cap B = \varnothing$), they are called disjoint.

The *difference of two sets* A and B, denoted by $A - B$, is the set of elements that belong to A but not to B. Thus:

$$A - B = \{x \mid x \in A \text{ and } x \notin B\}$$

For example, if $A = \{1, 2, 3, 4\}$ and $B = \{4, 5, 7\}$, then $A - B$ is equal to

$$A - B = \{1, 2, 3\}$$

When considering a collection of sets, say $A = \{2, 3, 7\}$ and $B = \{3, 11, 12\}$, it is useful to think of each set in the collection as a subset of some set U called the universal set. The choice of U depends only on the particular problem at hand. Let us say

$$U = \{1, 2, 3, 4, 5, 6, 7, 8, 9, 10, 11, 12\}$$

then with A and B as before, we have

$$U - A = \{1, 4, 5, 6, 8, 9, 10, 11, 12\}$$

and

$$U - B = \{1, 2, 4, 5, 6, 7, 8, 9, 10\}$$

The set $(U - A)$ is called the complement of A (in U) and the set $(U - B)$ is called the complement of B (in U). There are many ways to denote the complement of A. We will use $\bar{A}$ to denote the complement of A.

The power set of a set A is defined as the set of all subsets of A and is denoted as $P(A)$.

Example 1.3 Find the power set of $A = \{1, 2\}$.

Solution: The power set of A is $P(A) = \{\varnothing, \{1\}, \{2\}, \{1, 2\}\}$.

1.1.1 Venn diagrams

Unions, intersections and complements of sets can sometimes be visualized using so-called Venn diagrams. In Figure 1.2, the points on and inside the circle form set A and the points on and inside the oval form set B. The intersection of A and B consists of the shaded area. The points in the shaded area belong to both A and B.

1.1.2 Cartesian product

The Cartesian product of two sets A and B, denoted by $A \times B$, is the set of all possible ordered pairs whose first component is a member of A and whose second component is a member of B. Thus:

$$A \times B = \{(x, y) \mid x \in A \text{ and } y \in B\}$$

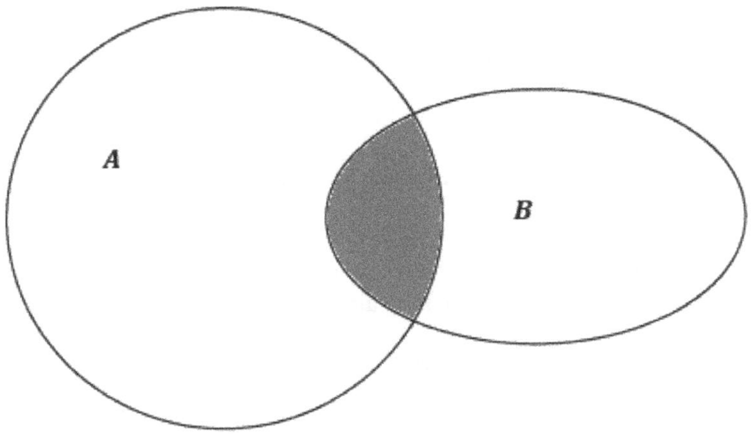

Figure 1.2

For example, if $A = \{2, 3\}$ and $B = \{4, 5, 6\}$, then

$$A \times B = \{(2, 4), (2, 5), (2, 6), (3, 4), (3, 5), (3, 6)\}$$

Example 1.4 Let $A = B = \mathbb{R}$. Then $A \times B = \mathbb{R} \times \mathbb{R} = \mathbb{R}^2$. The set $\mathbb{R}^2$ is often referred to as the real plane.

1.1.3 Relations

Let A and B be sets. A relation, R, from A to B is a subset of $(A \times B)$. Now, suppose $x \in A$ and $y \in B$. The notations xRy (x is in relation R to y) and $(x, y) \in R$ are equivalent. If $A = B$, then a relation R from A to B is known as a relation on A.

Example 1.5 Let $A = B = \mathbb{R}$. Define a relation R from A to B as follows: xRy if $x = y$. Then $(x, y) \in R$ implies that $x = y$.

1.1.4 Elementary types of relations on X

Let us consider a set X and a relation R on set X. A relation R on set X is said to be

(a) reflexive if xRx,
(b) symmetric if $xRy \Rightarrow yRx$,
(c) transitive if (xRy and yRz) $\Rightarrow xRz$,
(d) complete if either xRy or yRx,

for all $x, y, z \in X$.

A relation on X may have any combination of these four properties, including none.

Example 1.6 Consider the strict "less than" relation R_1 on a set of natural numbers $\mathbb{N}$. R_1 is not reflexive, since $n < n$ is not true. The relation R_1 is not symmetric, since $n < m$

does not imply $m < n$ for $n, m \in \mathbb{N}$. However, R_1 is transitive since $n < m$ and $m < w$ implies $n < w$ for all $n, m, w \in \mathbb{N}$.

EXAMPLE 1.7 Let $X = Y = \mathbb{R}$. The set of ordered pairs

$$\{(x, y) \in \mathbb{R}^2 \mid y = 3x\}$$

is a relation on $\mathbb{R}$, and its graphical counterpart is the set of points lying on the straight line $y = 3x$.

Observe that, given $y_1 \neq y_2$, if $(x_1, y_1) \in R$, then $(x_1, y_2) \notin R$. For instance, if ordered pair $(2, 6)$ belongs to the relationship, no other ordered pair with 2 as the first element belongs to the relationship.

A *function* f from a set A to a set B is a relation (from A to B) that assigns to each element in A at most one element in B. In this case, we write $f\colon A \to B$.

EXAMPLE 1.8 Consider a function f from the set of natural numbers $\mathbb{N}$ to the set of real numbers $\mathbb{R}$ given by $f(x) = \sqrt{x}$. Here $f(4) = 2$ but $f(4.5)$ is undefined, since 4.5 is not a natural number.

Now, if $f\colon A \to B$, the subset of A on which f is defined is called the *domain* of f; and the subset of B in which f takes its values is called the *range* of f.

EXAMPLE 1.9 Consider a function $f\colon \mathbb{R} \to \mathbb{R}$ defined by $f(x) = x^2$. Then, the domain of f is all of $\mathbb{R}$, written

$$\text{Domain}(f) = \mathbb{R}$$

and the range of f is the set of all non-negative real numbers, written

$$\text{Range}(f) = \{y \in \mathbb{R} \mid y \geq 0\}$$

1.1.5 Necessary and sufficient conditions

To better understand when certain mathematical methods can be used in applied situations, it is important to remember that every method has its limitations. The conditions that must be satisfied in order for us to use a certain formula are usually stated in a mathematical statement known as a theorem. To read theorems properly, we will remind the reader about the distinction between necessary and sufficient conditions.

Consider the statement: "It is necessary to graduate from a high school to enter a Ph.D. program." The statement means that, unless you have graduated from a high school, you cannot enter a Ph.D. program. It is a necessary condition. However, it is by no means sufficient to graduate from high school to be accepted into a Ph.D. program. The statement does *not* say that, if you are a high-school graduate, then you can enter a Ph.D. program.

Now consider the statement: "If n is a natural number, then n is a real number." The statement is equivalent to saying the following: in order for a number to be real, it is sufficient for it to be natural. By no means does it require the number to be natural: as we know there are real numbers that are not natural. This is an example of a sufficient condition.

If from a logical statement p we can deduce a logical statement q, then we will express this fact as $p \Rightarrow q$. We say "if p then q" or "q follows from p." If in addition p follows from q, then the statements p and q are logically equivalent and we write $p \Leftrightarrow q$. The logical equivalence statement $\Leftrightarrow$ means that p happens if and only if q does. Quite often an abbreviation "if and only if" is used instead of saying "if and only if." Sometimes the symbol $\Leftrightarrow$ is used.

Many times in the text we make statements valid for any object in a certain class, for instance every real number. The symbol $\forall$ is used as a short-cut for "for any." For instance,

$$\forall\, x > 0$$

reads "for any $x > 0$."

PROBLEMS FOR SECTION 1.1

1. Write the following in set notation.

   (a) the set of all real numbers
   (b) the set of all natural numbers
   (c) the set of all non-negative integers
   (d) the set of all real numbers greater than 76
   (e) the set of all natural numbers less than 22
   (f) the set that contains no elements

2. Let $A = \{1, 2, 3\}$, $B = \{3, 4, 7\}$, $C = \{3, 4\}$ and $D = \{7\}$.

   (a) Determine which of the following statements are true.

      (i) $5 \in A$
      (ii) $7 \in B$
      (iii) $D \subset B$
      (iv) $B \subset D$
      (v) $A \subset B$

   (b) Find the following sets.

      (i) $A \cap B$
      (ii) $A \cup B$
      (iii) $A - B$
      (iv) $B - A$
      (v) $(A \cap B) - A$
      (vi) $(B \cup A) - B$
      (vii) $(A \cup B) - (C \cup D)$
      (viii) $A \cap B \cap C \cap D$
      (ix) $A \cap C \cap D \cap \varnothing$
      (x) $((A \cup B) - C) - D$

3. Describe the power set of each of the following.

   (a) $B_1 = \{a\}$
   (b) $B_2 = \{a, b\}$
   (c) $B_3 = \{1, 2, 3\}$

(d) $B_4 = \{5, 7, 9, 11\}$
(e) $B_5 = \{\{a, b\}, \{a, c\}, \{b, c\}\}$

4. Which of the following statements are valid?

(a) $A \cup A = A$
(b) $A \cup \varnothing = A$
(c) $A \cup U = A$
(d) $A \cap A = A$
(e) $A \cap \varnothing = U$
(f) $A \cap U = U$
(g) $\bar{A} = A$

5. Prove that $A \subset B$ and $B \subset A$ imply that $A = B$.
6. A survey revealed that 80 students liked economics and 70 students liked finance. Further, 20 students liked both economics and finance, and 50 students liked neither. How many students responded to the survey?

1.2 Functions from $\mathbb{R}$ to $\mathbb{R}$

1.2.1 Polynomial functions

Most likely the simplest function that you will encounter is the constant function $f(x) = a$, where a is a constant. For example, $f(x) = 17.5$ (or alternatively $y = 17.5$) is a constant function. The value of the function y is always equal to 17.5 regardless of the value of x (see Figure 1.3). In this case, we have Domain$(f) = \mathbb{R}$ and Range$(f) = \{17.5\}$.

A constant function is a special case of a polynomial function. The general form of a polynomial function is

$$f(x) = \sum_{i=0}^{n} a_i x^i$$

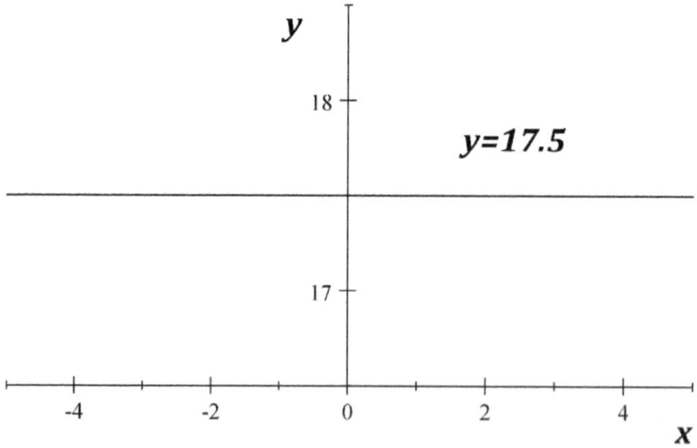

Figure 1.3

Fixing $n=0$, we have $y=a_0$. Letting $n=1,2,3,4,$ in turn, we have

$$y=a_0+a_1x \qquad \text{linear}$$

$$y=a_0+a_1x+a_2x^2 \qquad \text{quadratic}$$

$$y=a_0+a_1x+a_2x^2+a_3x^3 \qquad \text{cubic}$$

$$y=a_0+a_1x+a_2x^2+a_3x^3+a_4x^4 \qquad \text{quartic}$$

Linear functions yield graphs that are straight lines, hence the name linear. Recall that the equation of a linear function is given by $y=a_0+a_1x$, where a_0 is the intercept and a_1 is the slope. Examples of linear functions include the following:

$$y_1=2x+5$$

$$y_2=-3x+7$$

$$y_3=7x$$

The functions above are represented graphically in Figure 1.4.

1.2.2 Zeros of quadratic functions

DEFINITION 1.1 The zeros of a function are the x-values where the graph of the function intersects the x-axis. Algebraically, zeros of function $f(x)$ are numbers x_i such that $f(x_i)=0$.

A polynomial function of degree n has n zeros, provided multiple and complex zeros are counted. The zeros of polynomial functions are also known as roots.

A quadratic function is given by the following equation:

$$y=a_0+a_1x+a_2x^2, \quad \text{where } a_2 \neq 0$$

Figure 1.5 depicts the simplest ($a_0=0,\ a_1=0,\ a_2=1$) quadratic function $y=x^2$.

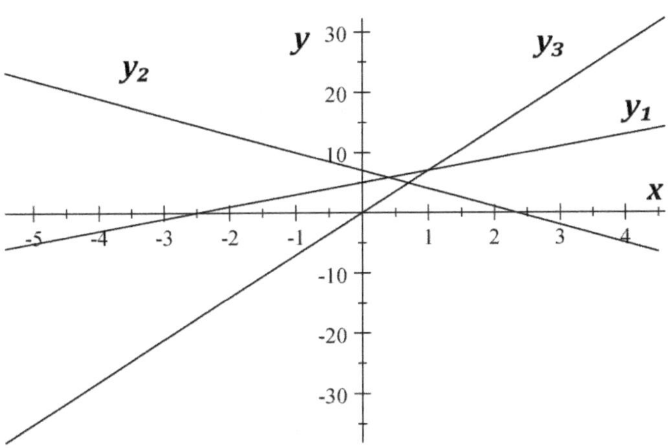

Figure 1.4

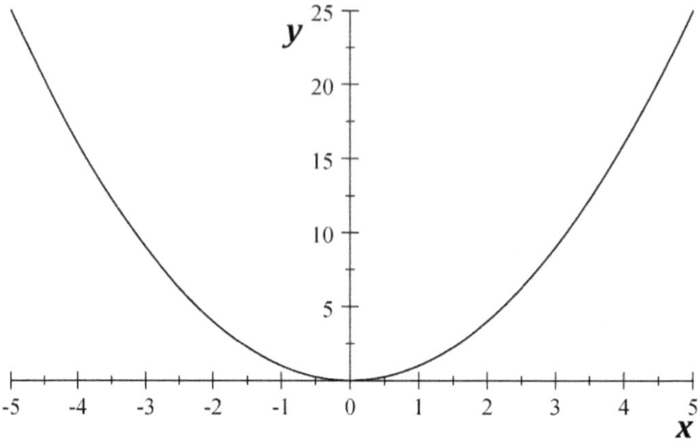

Figure 1.5 $y = x^2$.

The graph of a quadratic function is called a parabola. A parabola is U-shaped and may open up or down. Now, let us generalize slightly and consider the function

$$y = ax^2$$

for $a = 1, 3$ and 5 (see Figure 1.6).

A parabola ($y = a_0 + a_1x + a_2x^2$) opens up and has a minimum when $a_2 > 0$ and opens down and has a maximum when $a_2 < 0$.

Probably, the most common form of representing the quadratic function is

$$y = f(x) = ax^2 + bx + c$$

We use this representation to find the vertex and the roots of quadratic functions.

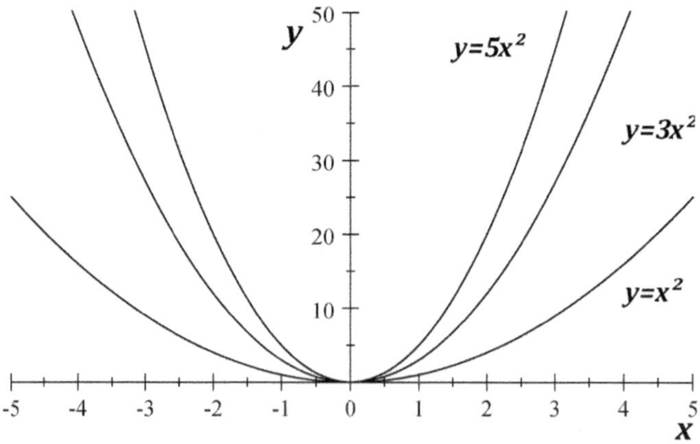

Figure 1.6

The maximum/minimum point of a parabola is called the *vertex* of the parabola. The coordinates of the vertex are given by

$$(x, y) = \left(-\frac{b}{2a}, f\left(-\frac{b}{2a}\right)\right)$$

EXAMPLE 1.10 Let us consider the function $f(x) = x^2 - 4$. The graph of the function is given in Figure 1.7.

We can find the zeros of this function both graphically and analytically. Figure 1.7 shows that the function $y = x^2 - 4$ has two zeros at $x_1 = -2$ and $x_2 = 2$. Indeed, $f(-2) = f(2) = 0$. The coordinates of the vertex are

$$\left(-\frac{b}{2a}, f\left(-\frac{b}{2a}\right)\right) = (0, -4).$$

We can write the formula for $f(x)$ in the following way:

$$f(x) = x^2 - 4 = (x+2)(x-2) \tag{1.1}$$

It is seen rather readily from (1.1) that the zeros of the function are $x_1 = -2$ and $x_2 = 2$.

In general, the zeros of the quadratic function $y = ax^2 + bx + c$ are found using the formula:

$$x_{1,2} = \frac{-b \pm \sqrt{b^2 - 4ac}}{2a}$$

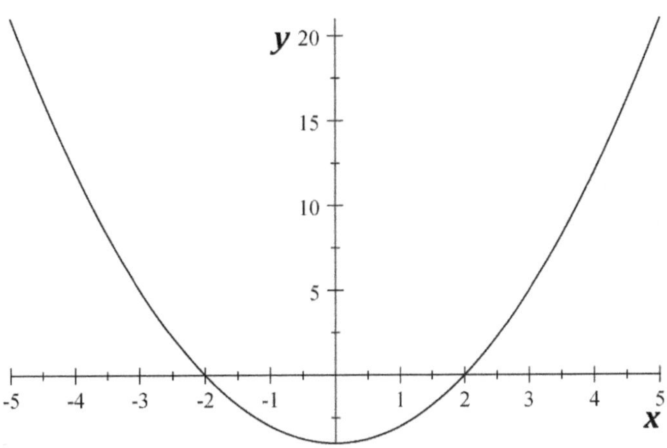

Figure 1.7 $y = x^2 - 4$.

There are three cases to consider.
Case 1. If $b^2 - 4ac < 0$, then $f(x) = ax^2 + bx + c$ has no real zeros.
Case 2. If $b^2 - 4ac = 0$, then $f(x) = ax^2 + bx + c$ has one real zero, namely,

$$x_1 = -\frac{b}{2a}$$

Case 3. If $b^2 - 4ac > 0$, then $f(x) = ax^2 + bx + c$ has two distinct zeros,

$$x_1 = \frac{-b + \sqrt{b^2 - 4ac}}{2a} \quad \text{and} \quad x_2 = \frac{-b - \sqrt{b^2 - 4ac}}{2a}$$

EXAMPLE 1.11 Suppose a firm produces and sells cars. The price P per unit is

$$P(Q) = 20 - 3Q$$

and the cost of producing Q units is

$$C(Q) = 4Q + Q^2$$

Find the profit-maximizing quantity Q^* and the maximal profit π^*.

Solution: Note that

$$\pi(Q) = \text{TR} - \text{TC} = PQ - C = 16Q - 4Q^2$$

where TR is total revenue and TC is total cost. It follows that the profit is maximized at

$$Q^* = -\frac{16}{2(-4)} = 2$$

with

$$\pi^* = \pi(Q^*) = 16 \cdot 2 - 4 \cdot 2^2 = 16$$

PROBLEMS FOR SECTION 1.2

1. Find the zeros of the following quadratic functions.

(a) $y = x^2 - 7x + 10$
(b) $y = x^2 + 8x + 15$
(c) $y = x^2 - 9x + 9$
(d) $y = x^2 - 6x - 16$
(e) $y = x^2 - 2x - 15$
(f) $y = 2x^2 - 17x - 30$
(g) $y = 4x^2 - 12x + 9$

2. Find the maximum/minimum of the following quadratic functions.

   (a) $y = x^2 - 5x + 10$
   (b) $y = x^2 + 4x + 1$
   (c) $y = x^2 - 9x + 9$
   (d) $y = x^2 - 6x$
   (e) $y = x^2$
   (f) $y = 2x^2 - 14x - 30$
   (g) $y = 3x^2 - x + 7$

2 Fundamental functions and series

Most concepts that economists are concerned with are quantitative in nature. Therefore, much of economic analysis is invariably mathematical in nature. To express relationships between economic variables, economists employ various mathematical functions. The previous chapter introduced the reader to the set of polynomial functions. The first section of this chapter focuses on various types of power functions. The following two sections explore finite and infinite sequences and series, and offer some rules of summation. Economics students will utilize this knowledge in their econometrics and time-series analysis courses.

2.1 Power functions

Consider the general power function $f(x) = Ax^m$ $(x \in \mathbb{R}^+,\ m \in \mathbb{R}, A \in \mathbb{R})$. We have to assume that $x > 0$ because we want Ax^m to be defined for all values of x and m. Observe that Ax^m is undefined for $x = -2$ and $m = \frac{1}{4}$. In addition, since 0^0 is undefined, we require that $x \neq 0$.

Figure 2.1 illustrates the graph of $y = x^m$ for different values of m, namely $m = \frac{1}{4}$, $m = \frac{1}{2}$, $m = 1$ and $m = 2$. Since $f(1) = 1^m = 1$ for any m, we conclude that $(x, y) = (1, 1)$ may be called a special point for power functions as it belongs to any power function x^m.

Figure 2.2 illustrates the graph of $y = x^m$ for different negative values of m, namely $m = -1$, $m = -2$ and $m = -4$. Note that the point (1, 1) belongs to all of the graphs.

2.2 Exponents

Let n be a natural number. We define

$$x^n = x \cdot x \cdot x \cdot \cdots \cdot x \ (n \text{ times})$$

EXAMPLE 2.1

(a) $5^4 = 5 \cdot 5 \cdot 5 \cdot 5 = 625$
(b) $3^1 = 3$
(c) $12^3 = 12 \cdot 12 \cdot 12 = 1728$

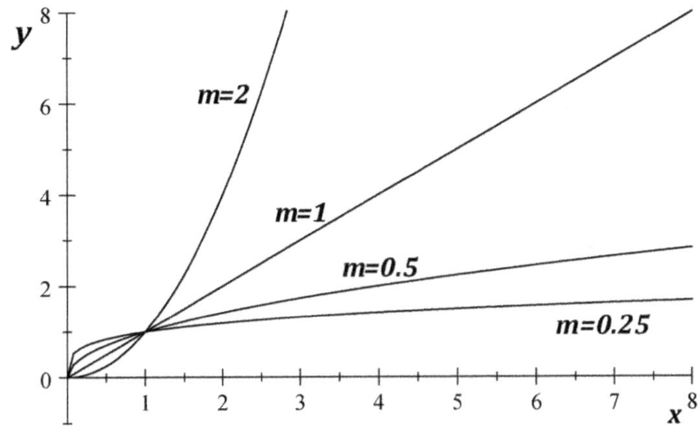

Figure 2.1

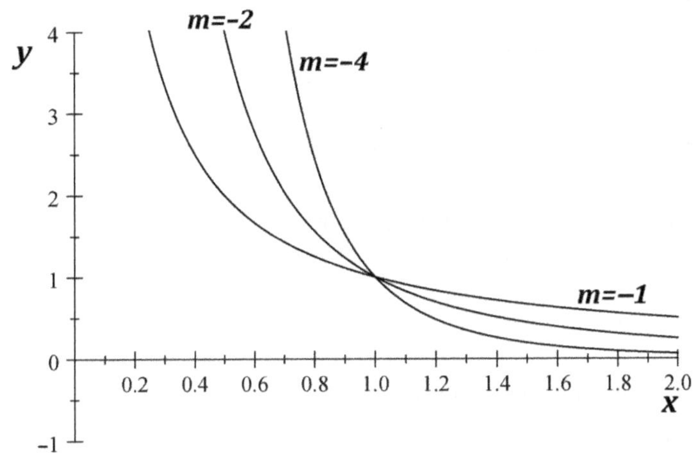

Figure 2.2

Note the following widely used rules for exponents: if $m, n \in \mathbb{Z}$ then

(Exp1) $x^m \cdot x^n = x^{m+n}$

(Exp2) $(x^m)^n = x^{m \cdot n}$

(Exp3) $(xy)^n = x^n y^n$

(Exp4) $\left(\dfrac{x}{y}\right)^n = \dfrac{x^n}{y^n}$

(Exp5) $\dfrac{x^m}{x^n} = x^{m-n} = x^m \cdot x^{-n}$

Observe that **Exp5** assumes that $y \neq 0$. Now, using **Exp1–Exp5** we have

(a) $4^7 \cdot 4^{11} = 4^{7+11} = 4^{18}$

(b) $(5^3)^8 = 5^{3 \cdot 8} = 5^{24}$

(c) $(7 \cdot 11)^4 = 7^4 \cdot 11^4$

(d) $\left(\frac{3}{5}\right)^{17} = \frac{3^{17}}{5^{17}}$

DEFINITION 2.1 If n is an integer, and $x \neq 0$, then

$$x^{-n} = \frac{1}{x^n} \tag{2.1}$$

It follows from equation (2.1) that

(a) $3^{-3} = \frac{1}{3^3} = \frac{1}{27}$

(b) $5^{-2} = \frac{1}{5^2} = \frac{1}{25}$

EXAMPLE 2.2 Compute $\left(\frac{2}{3}\right)^{-5}$.

Solution: Following the explanation above we get

$$\left(\frac{2}{3}\right)^{-5} = \frac{1}{\left(\frac{2}{3}\right)^5} = \left(\frac{3}{2}\right)^5$$

It follows readily that if a, b, c, d and e are constants, then

$$\frac{x^a \cdot x^b \cdot x^c}{x^d \cdot x^e} = x^{a+b+c-d-e}$$

This rule can be extended to any number of constants:

$$\frac{x^{K_1} \cdot x^{K_2} \cdot \dots \cdot x^{K_j}}{x^{K_{j+1}} \cdot x^{K_{j+2}} \cdot \dots \cdot x^{K_w}} = x^{K_1+K_2+\dots+K_j-K_{j+1}-K_{j+2}-\dots-K_w}$$

EXAMPLE 2.3 Simplify the following expressions.

(a) $\frac{x^3 \cdot x^7 \cdot x^{11}}{x^2 \cdot x^5 \cdot x^{14} \cdot x^6}$

(b) $\frac{x^3 \cdot y^4 \cdot z^5}{y^2 \cdot x^7 \cdot z^3}$

Solution:

(a) $\frac{x^3 \cdot x^7 \cdot x^{11}}{x^2 \cdot x^5 \cdot x^{14} \cdot x^6} = x^{3+7+11-2-5-14-6} = x^{-6}$

(b) $\frac{x^3 \cdot y^4 \cdot z^5}{y^2 \cdot x^7 \cdot z^3} = x^{3-7}y^{4-2}z^{5-3} = x^{-4}y^2z^2$

2.2.1 Higher roots: *n*th roots

We have defined x^n, where n is a natural number, as x multiplied by itself n times. What if n is a rational number, not an integer, such as $\frac{1}{2}$? It turns out that x^n can be defined for $n = \frac{1}{2}$ and $x \geq 0$ as follows: $x^{\frac{1}{2}} = \sqrt[2]{x}$.

In general, $x^{\frac{1}{m}}$, where m is a natural number, is defined as follows:

$$x^{\frac{1}{m}} = \sqrt[m]{x}$$

where x must be non-negative if m is even and may be any real number if m is odd.

EXAMPLE 2.4 Compute the following.

(a) $32^{\frac{1}{5}}$
(b) $\left(\frac{1}{16}\right)^{\frac{1}{4}}$
(c) $(-0.125)^{\frac{1}{3}}$

Solution:

(a) $32^{\frac{1}{5}} = (2^5)^{\frac{1}{5}} = 2^1 = 2$
(b) $\left(\frac{1}{16}\right)^{\frac{1}{4}} = \left(\frac{1}{2^4}\right)^{\frac{1}{4}} = (2^{-4})^{\frac{1}{4}} = 2^{-1} = \frac{1}{2}$
(c) $(-0.125)^{\frac{1}{3}} = \left(-\frac{1}{8}\right)^{\frac{1}{3}} = ((-2)^{-3})^{\frac{1}{3}} = (-2)^{-1} = -\frac{1}{2}$

It is very important to understand that 2^{-1} is not a negative number even though 2 is raised to a negative power. Recall that 2^{-1} equals $\frac{1}{2}$, which is positive. Further, -2^{-1} is not a positive number and the two negative signs do not cancel out. In fact, -2^{-1} is a negative number: $-2^{-1} = -\frac{1}{2} < 0$. We could also write -2^{-1} as $(-1)(2^{-1}) = -\frac{1}{2}$.

EXAMPLE 2.5 Compute the following.

(a) -3^2
(b) $(-3)^2$
(c) -3^3
(d) $(-3)^3$

Solution:

(a) $-3^2 = (-1)(3^2) = (-1)9 = -9$
(b) $(-3)^2 = (-3)(-3) = 9$
(c) $-3^3 = (-1)3^3 = (-1)27 = -27$
(d) $(-3)^3 = (-3)(-3)(-3) = -27$

It follows from (c) and (d) that $(-3)^3 = -3^3$. Indeed,

$$(-3)^3 = ((-1) \cdot 3)^3 = -27 = (-1)^3 \cdot 3^3 = -3^3$$

Now, we are ready to define $x^{m/n}$, where m and n are natural numbers:

$$x^{m/n} = \sqrt[n]{x^m}$$

As earlier, x must be non-negative if n is even. Rules **R1–R5** in fact apply to all real exponents.

EXAMPLE 2.6 Compute the following.

(a) $4^{\frac{3}{2}}$
(b) $(-8)^{\frac{2}{3}}$

Solution:

(a) $4^{\frac{3}{2}} = \sqrt[2]{4^3} = \sqrt[2]{64} = 8$
(b) $(-8)^{\frac{2}{3}} = \sqrt[3]{(-8)^2} = \sqrt[3]{64} = 4$

PROBLEMS FOR SECTION 2.2

1. Compute the following without using a calculator.

   (a) $64^{\frac{1}{2}}$
   (b) $81^{\frac{1}{3}}$
   (c) $32^{\frac{1}{5}}$
   (d) $4^{\frac{3}{2}}$
   (e) $8^{\frac{2}{3}}$
   (f) $10{,}000^{\frac{1}{4}}$
   (g) $0.125^{-\frac{1}{3}}$
   (h) $0.25^{-\frac{1}{2}}$

2. Simplify the following.

   (a) $(3x^2y^{\frac{1}{4}}z^5)^8$
   (b) $(w^a m^b k^{3t})^{a/b}$
   (c) $\dfrac{(x+w+e+r)^{\frac{17}{2}}}{(x+w+e+r)^{\frac{15}{2}}}$
   (d) $\dfrac{(a+b+c)^{12}(d+e)^{11}}{(a+b+c)^5(d+e)^{12}}$
   (e) $\left(\dfrac{(a+b+c)^{12}(d+e)^{11}}{(a+b+c)^5(d+e)^{12}}\right)^0$
   (f) $\dfrac{a^{12}b^3c^5d^{31}e^{32}f^{11}}{d^4e^{12}c^{31}a^{23}b^{12}}$

(g) $\dfrac{\sqrt[13]{5}\,\sqrt[15]{7}\,\sqrt[17]{11}}{\sqrt[5]{5}\,\sqrt[7]{7}\,\sqrt[15]{11}}$

(h) $\sqrt[3]{2}\,\sqrt[3]{2}\,\sqrt[3]{2}$

(i) $\dfrac{\sqrt[4]{5}\,\sqrt[6]{5}\,\sqrt[8]{5}}{\sqrt[2]{5}}$

(j) $\dfrac{\sqrt[2]{2.5}\,\sqrt[2]{12}}{\sqrt[3]{2}\,\sqrt[3]{13.5}}$

2.3 Sequences and series

A *sequence* is a function whose domain is a subset of the set of natural numbers, $\mathbb{N}$. For example,

$$f(x) = \frac{1}{x}, \quad \text{where } x \in \mathbb{N}$$

is a sequence. A sequence is usually represented by listing its values in order:

$$f(1) = 1$$
$$f(2) = \frac{1}{2}$$
$$f(3) = \frac{1}{3}$$

and so on. More often, a sequence is represented as an ordered list:

$$1, \frac{1}{2}, \frac{1}{3}, \frac{1}{4}, \ldots$$

The ellipsis dots "..." indicate that the sequence

$$f(x) = \frac{1}{x}, \quad \text{where } x \in \mathbb{N}$$

never ends. The numbers in this list are called the terms of the sequence.

Usually we use subscripted (indexed) letters when working with sequences. For instance,

$$a_1 = 1, \quad a_2 = \frac{1}{2}, \quad a_3 = \frac{1}{3}, \quad a_4 = \frac{1}{4}, \quad \ldots, \quad a_k = \frac{1}{k}, \quad \ldots$$

For the sequence above, we have a specific rule for the kth term of the sequence ($a_k = 1/k$) and therefore can easily find any term of the sequence. Whenever a formula for the kth term of a sequence is known, we usually represent the sequence by placing braces around the formula for the kth term. For example, the sequence

$$f(x) = \frac{1}{x}, \quad \text{where } x \in \mathbb{N}$$

may be represented as

$$\{a_k\} = \left\{\frac{1}{k}\right\}$$

EXAMPLE 2.7 Write down the first five terms of the following sequence:

$$\{b_k\} = \left\{(-1)^k \left(\frac{1}{k}\right)\right\}$$

Solution: The first five terms are

$$b_1 = -1, \quad b_2 = \frac{1}{2}, \quad b_3 = -\frac{1}{3}, \quad b_4 = \frac{1}{4}, \quad b_5 = -\frac{1}{5}$$

EXAMPLE 2.8 Write down the first five terms of the following sequence:

$$\{b_k\} = \left\{(-1)^{k+1} (k+2)^2\right\}$$

Solution: The first five terms are

$$b_1 = 9, \quad b_2 = -16, \quad b_3 = 25, \quad b_4 = -36, \quad b_5 = 49$$

Now, let us consider the sequence

$$\{a_k\} = \left\{\frac{1}{k}\right\}$$

The first five elements of this sequence are

$$a_1 = 1, \quad a_2 = \frac{1}{2}, \quad a_3 = \frac{1}{3}, \quad a_4 = \frac{1}{4}, \quad a_5 = \frac{1}{5}$$

Note that

$$a_1 > a_2 > a_3 > a_4 > a_5$$

for this sequence. Generally, sequences for which

$$a_n > a_{n+1} \quad \text{for } any\ n \in \mathbb{N}$$

are called *decreasing.*

EXAMPLE 2.9 Verify that the sequence

$$\{b_k\} = \left\{\left(\frac{1}{k}\right)^2\right\}$$

is decreasing.

Solution: It is sufficient to show that

$$b_k > b_{k+1} \quad \forall\, k \in \mathbb{N}$$

to prove that the sequence $\{b_k\}$ is decreasing. Since

$$\left(\frac{1}{k}\right)^2 > \left(\frac{1}{k+1}\right)^2 \quad \forall\, k \in \mathbb{N}$$

we conclude that the sequence $\{b_k\}$ is decreasing.

EXAMPLE 2.10 Verify that the sequence

$$\{b_n\} = \left\{\left(\frac{1}{n}\right) + \left(\frac{1}{n+1}\right)\right\}$$

is decreasing.

Solution: Again, it is sufficient to show that

$$b_n > b_{n+1} \quad \forall\, n \in \mathbb{N}$$

to prove that the sequence $\{b_n\}$ is decreasing. Since

$$\frac{1}{n} + \frac{1}{n+1} > \frac{1}{n+1} + \frac{1}{n+2} \quad \forall\, n \in \mathbb{N}$$

we conclude that the sequence $\{b_n\}$ is decreasing.

Now, let us consider the sequence

$$\{a_k\} = \left\{k^2\right\}$$

The first five elements of this sequence are

$$a_1 = 1, \quad a_2 = 4, \quad a_3 = 9, \quad a_4 = 16, \quad a_5 = 25$$

Note that

$$a_1 < a_2 < a_3 < a_4 < a_5$$

for this sequence. Sequences for which

$$a_k < a_{k+1} \quad \text{for } any\ k \in \mathbb{N}$$

are called *increasing*.

EXAMPLE 2.11 Verify that the sequence

$$\{b_n\} = \left\{n^3\right\}$$

is increasing.

Solution: It is sufficient to show that

$$b_n < b_{n+1} \quad \forall\, n \in \mathbb{N}$$

to prove that the sequence $\{b_n\}$ is increasing. Since

$$n^3 < (n+1)^3 \quad \forall\, n \in \mathbb{N}$$

we conclude that the sequence $\{b_n\}$ is decreasing.

EXAMPLE 2.12 Verify that the sequence

$$\{b_n\} = \left\{\frac{n+1}{n}\right\}$$

is increasing.

Solution: Again, it is sufficient to show that

$$b_n < b_{n+1} \quad \forall\, n \in \mathbb{N}$$

to prove that the sequence $\{b_n\}$ is increasing. Since

$$\frac{n+1}{n} > \frac{n+2}{n+1} \quad \forall\, n \in \mathbb{N}$$

we conclude that the sequence $\{b_n\}$ is increasing.

A sequence is called *monotonic* if it is strictly increasing or strictly decreasing. You may think that every sequence is monotonic; however, there are many sequences that are not. For example, consider the following sequence:

$$\{w_k\} = \{17\}$$

Every term of this sequence is equal to 17, so the sequence is neither strictly increasing nor strictly decreasing and therefore is not monotonic. Incidentally, every element of $\{w_k\}$ is less than 18 and more than 16. So, informally speaking, the sequence is bounded above by 18 and bounded below by 16. Let us consider the following definitions.

DEFINITION 2.2 Sequence $\{w_k\}$ is *bounded above,* if and only if there exists a number M such that

$$b_n \leq M \quad \forall\, n \in \mathbb{N}$$

The number M is called an *upper bound*.

Furthermore, sequence $\{w_k\}$ is *bounded below*, if and only if there exists a number m such that

$$b_n \geq m \quad \forall\, n \in N$$

The number m is called a *lower bound*.

A sequence is said to be *bounded* if it is bounded above and below.

EXAMPLE 2.13 The sequence

$$\{b_k\} = \left\{\frac{1}{k}\right\}$$

is bounded below by 0 and above by 1. So the sequence $\{b_k\}$ is bounded. Its lower bound is 0 and its upper bound is 1.

EXAMPLE 2.14 The sequence

$$\{b_k\} = \{k\}$$

is bounded below by 0 and is not bounded above. So the sequence $\{b_k\}$ is unbounded.

A sequence may be infinite or finite. If a sequence has a terminal value

$$a_1, a_2, \ldots, a_n$$

it is called a finite sequence and the values $a_1, a_2, \ldots, a_n$ are its terms. However, if a sequence does not terminate, as in

$$\{b_k\} = \left\{\frac{1}{k}\right\} \quad \text{where } k \in \mathbb{N}$$

it is called infinite. The sum of the terms of a sequence is called a series:

$$S_n = a_1 + a_2 + \cdots + a_n$$

The series S_n is a finite series since it is a sum of a finite sequence. We often use $\sum$(sigma) for the summation:

$$S_n = \sum_{i=1}^{n} a_i$$

Similarly, the sum of an infinite sequence will produce an infinite series:

$$S = \sum_{i=1}^{\infty} a_i$$

We do not have to use index i to label the elements in our series. Also, the number n meant the number of elements in our series and can be denoted by something else if convenient. The letters j, k, l, m, n, p and r can be used, among others. For example, we can write

$$S_n = \sum_{j=1}^{n} a_j$$

$$S_n = \sum_{k=1}^{n} a_k$$

$$S_n = \sum_{p=1}^{n} a_p$$

to indicate the same thing. Observe that

$$\sum_{i=1}^{n} a_i = \sum_{k=1}^{n} a_k = \sum_{p=1}^{n} a_p = S_n$$

EXAMPLE 2.15 Write out the following in full.

$$\sum_{i=3}^{7} a_i$$

Solution:

$$\sum_{i=3}^{7} a_i = a_3 + a_4 + a_5 + a_6 + a_7$$

EXAMPLE 2.16 Write out the following in full.

$$\sum_{i=3}^{5} (a_i + i)$$

Solution:

$$\sum_{i=3}^{5} (a_i + i) = (a_3 + 3) + (a_4 + 4) + (a_5 + 5)$$

EXAMPLE 2.17 Write out the following in full.

$$\sum_{i=3}^{5} (a_i + i)^3$$

Solution:

$$\sum_{i=3}^{5} (a_i + i)^3 = (a_3 + 3)^3 + (a_4 + 4)^3 + (a_5 + 5)^3$$

Example 2.18 Write out the following in full.

$$\sum_{i=3}^{5}(a_i+i)^{2i+5}$$

Solution:

$$\sum_{i=3}^{5}a_i=(a_3+3)^{11}+(a_4+4)^{13}+(a_5+5)^{15}$$

2.4 Some rules of summation

Let us discuss some rules of summation. The first rule relates to the case where each element of the sum is equal to the same fixed number c, usually referred to as a constant.

(Sum1) If c is a constant, then

$$\sum_{i=1}^{n}c=nc$$

Example 2.19 Evaluate the following sum.

$$\sum_{i=1}^{5}17$$

Solution: According to **Sum1** we have

$$\sum_{i=1}^{5}17=5\cdot 17=85$$

We will use the following rules of summation on a regular basis.

(**Sum2**) If c is a constant, then

$$\sum_{i=1}^{n}cx_i=c\sum_{i=1}^{n}x_i$$

Proof:

$$\begin{aligned}\sum_{i=1}^{n}cx_i&=cx_1+cx_2+\cdots+cx_n\\&=c(x_1+x_2+\cdots+x_n)\\&=c\sum_{i=1}^{n}x_i\end{aligned}$$

(**Sum3**) If $\{x_i\}$ and $\{y_i\}$ are two sequences, then

$$\sum_{i=1}^{n}(x_i + y_i) = \sum_{i=1}^{n} x_i + \sum_{i=1}^{n} y_i$$

Proof:

$$\begin{aligned}\sum_{i=1}^{n}(x_i + y_i) &= (x_1 + y_1) + (x_2 + y_2) + \cdots + (x_n + y_n)\\ &= (x_1 + x_2 + \cdots + x_n) + (y_1 + y_2 + \cdots + y_n)\\ &= \sum_{i=1}^{n} x_i + \sum_{i=1}^{n} y_i\end{aligned}$$

(**Sum4**) If $\{x_i\}$ and $\{y_i\}$ are two sequences and c_1 and c_2 are constants, then

$$\sum_{i=1}^{n}(c_1 x_i + c_2 y_i) = c_1 \sum_{i=1}^{n} x_i + c_2 \sum_{i=1}^{n} y_i$$

Proof:

$$\begin{aligned}\sum_{i=1}^{n}(c_1 x_i + c_2 y_i) &= (c_1 x_1 + c_2 y_1) + (c_1 x_2 + c_2 y_2) + \cdots + (c_1 x_n + c_2 y_n)\\ &= (c_1 x_1 + c_1 x_2 + \cdots + c_1 x_2) + (c_2 y_1 + c_2 y_2 + \cdots + c_2 y_n)\\ &= c_1 \sum_{i=1}^{n} x_i + c_2 \sum_{i=1}^{n} y_i\end{aligned}$$

Note, that **Sum4** can be extended to any number of constants and sequences. For example, let $c_1, c_2, \ldots, c_m$ be m constants and $\{x_i^1\}, \{x_i^2\}, \ldots, \{x_i^m\}$ be m sequences. Then

$$\begin{aligned}&\sum_{i=1}^{n}(c_1 x_i^1 + c_2 x_i^2 + \cdots + c_m x_i^m)\\ &\quad = c_1 \sum_{i=1}^{n} x_i^1 + c_2 \sum_{i=1}^{n} x_i^2 + \cdots + c_m \sum_{i=1}^{n} x_i^m\end{aligned}$$

One comment should be made in passing. Sometimes the range of summation is clear from the context. In such a case, some authors may omit the range of summation and write $\sum x_i$ instead of

$$\sum_{i=1}^{n} x_i$$

Also, keep in mind that there are several ways to express the same sum. For example,

$$\sum_{i=1}^{n} x_i = \sum_{i=0}^{n-1} x_{i+1} = \sum_{i=5}^{n+4} x_{i-4}$$

In the above example, although we have changed the initial and terminal values of summation, the sum remains constant.

EXAMPLE 2.20 Show that

$$\sum_{i=1}^{n}(x_i + 1)(x_i - 1) = \left(\sum_{i=1}^{n} x_i^2\right) - n$$

Solution:

$$\begin{aligned}\sum_{i=1}^{n}(x_i + 1)(x_i - 1) &= \sum_{i=1}^{n}(x_i^2 - 1)\\ &= \sum_{i=1}^{n} x_i^2 - \sum_{i=1}^{n} 1 = \left(\sum_{i=1}^{n} x_i^2\right) - n\end{aligned}$$

2.4.1 Finite double sums

Let us consider the following sum:

$$\begin{aligned}&A_{11} + A_{12} + A_{13} + A_{14} + A_{21} + A_{22} + A_{23} + A_{24}\\ &\quad + A_{31} + A_{32} + A_{33} + A_{34}\end{aligned}$$

We can write the above sum as

$$\sum_{i=1}^{4} A_{1i} + \sum_{i=1}^{4} A_{2i} + \sum_{i=1}^{4} A_{3i}$$

or, more compactly, as

$$\sum_{j=1}^{3}\sum_{i=1}^{4} A_{ji}$$

Since the order of summation is unimportant in sums, we can write

$$\sum_{j=1}^{3}\sum_{i=1}^{4} A_{ji} = \sum_{i=1}^{4}\sum_{j=1}^{3} A_{ji}$$

PROBLEMS FOR SECTION 2.4

1. Express the sums below in a summation notation.

   (a) $1+2+3+\cdots+99+100$

   (b) $\frac{1}{2}+\frac{1}{3}+\frac{1}{4}+\cdots+\frac{1}{100}$

   (c) $\frac{1}{4}+\frac{1}{9}+\frac{1}{16}+\frac{1}{25}+\cdots+\frac{1}{100}$

   (d) $1\cdot 2+2\cdot 3+3\cdot 4+\cdots+99\cdot 100$

   (e) $1+4+9+16+25+36+\cdots+121$

   (f) $w+\frac{w^3}{3}+\frac{w^5}{5}+\frac{w^7}{7}+\cdots+\frac{w^{35}}{35}$

   (g) $1-2+3-4+5-6+\cdots+99$

   (h) $1\cdot 4+2\cdot 5+3\cdot 6+4\cdot 7+\cdots+12\cdot 15$

   (i) $\frac{1\cdot 2}{3\cdot 4}+\frac{2\cdot 3}{4\cdot 5}+\frac{3\cdot 4}{5\cdot 6}+\cdots+\frac{96\cdot 97}{98\cdot 99}$

2. Expand the following sums.

   (a) $\sum_{i=1}^{3}\frac{i}{i+1}$

   (b) $\sum_{i=3}^{5}\left(\frac{1}{i^2}+\frac{1}{i^3}\right)$

   (c) $\sum_{i=1}^{4}(i+1)(i+2)(i+3)$

   (d) $\sum_{i=2}^{4}(i+1)^2(2i+1)$

   (e) $\sum_{i=-1}^{3}(i+1)(i+4)(i-3)$

   (f) $\sum_{i=1}^{3}(i+2i^2+3i^3)^2$

   (g) $\sum_{i=2}^{3}\left(\frac{1}{i+1}+\frac{4i}{i+3}\right)$

   (h) $\sum_{i=-3}^{-1}\left(\frac{1}{i+1}+\frac{i+1}{i+2}+\frac{(i+3)^{i+4}}{(i+5)^{i+6}}\right)$

3. Which of the following statements are true?

(a) $\sum_{i=1}^{n} a_i = \sum_{j=1}^{n} a_j = \sum_{k=1}^{n} a_k$

(b) $\sum_{i=1}^{10} (i+1)^2 = \sum_{i=1}^{7} (i+1)^2 + \sum_{i=8}^{10} (i+1)^2$

(c) $\sum_{i=2}^{8} 17i^2 = 17\left(\sum_{i=2}^{8} i^2\right)$

(d) $\sum_{i=1}^{5} \left(\frac{a_i+1}{b_i+1}\right) = \frac{\sum_{i=1}^{5}(a_i+1)}{\sum_{i=1}^{5}(b_i+1)}$

(e) $\sum_{i=1}^{5} (a+b+c)^3 = \sum_{i=1}^{5} a^3 + \sum_{i=1}^{5} b^3 + \sum_{i=1}^{5} c^3$

4. Expand the following sums.

(a) $\sum_{i=1}^{3}\sum_{j=2}^{4}(2i+3j)$

(b) $\sum_{i=1}^{3}\sum_{j=-1}^{2}(i^2+j^2)^{i+j}$

(c) $\sum_{i=1}^{3}\sum_{j=3}^{4}(i+j)(a_i+b_j)$

5. Prove that the following is true.

$$\sum_{i=1}^{3}\sum_{j=2}^{4}(i+j) = \sum_{j=2}^{4}\sum_{i=1}^{3}(j+i)$$

3 Exponential and logarithmic functions

Logarithmic functions are indispensable in economic analysis as they can transform multiplicative relationships between economic variables into additive ones. In addition, economists often choose (for the sake of convenience) to optimize the natural log of an objective function instead of the objective function itself. Exponential functions are very useful when modeling the growth of a certain economic variable. For example, we can use exponential functions to model the growth of the population of a country.

This chapter introduces logarithmic and exponential functions and closes with some Mathematica examples.

3.1 Logarithmic function

Consider the following equation:

$$3^5 = 243$$

In the equation above the base is equal to 3, and the exponent is equal to 5. The power to which 3 must be raised to yield 243 is called the logarithm (or log) to the base 3 of 243. So, logarithm to the base 3 of 243 is equal to 5:

$$\log_3 243 = 5$$

In general, if

$$a^x = B$$

where a, $B > 0$ and $a \neq 1$, and $x \in \mathbb{R}$, then we can write that

$$\log_a B = x$$

For example, $2^5 = 32$ implies that log to the base 2 of 32 equals 5

$$\log_2 32 = 5$$

EXAMPLE 3.1

(a) $\log_2 1 = 0$, since 2 raised to the power of 0 is 1.
(b) $\log_7 7 = 1$, since 7 raised to the power of 1 is 7.

(c) $\log_{10} \frac{1}{10} = -1$, since 10 raised to the power of -1 is $\frac{1}{10}$.
(d) $\log_3 81 = 4$, since 3 raised to the power of 4 is 81.
(e) $\log_5(-25)$ is not defined, since $-25 < 0$.
(f) $\log_{12} 1728 = 3$, since 12 raised to the power of 3 is 1728.
(g) $\log_5 \frac{1}{25} = -2$, since 5 needs to be raised to the power of -2 in order to get $\frac{1}{25}$.

Often economists work with logarithms to the base e, where e is the irrational number 2.718... known as the *exponential.* Logarithms to the base e are referred to as natural logarithms. We can write either $\log_e B$ or $\ln B$ to refer to the natural logarithm of B.

Now, consider the function $y = f(x) = \ln x$. The graph of y is given in Figure 3.1. The domain of $f(x)$ is all positive numbers,

$$\text{Domain}(f) = \{x \mid x > 0\}$$

and the range of $f(x)$ is all real numbers,

$$\text{Range}(f) = \mathbb{R}$$

The function $y = \ln x$ has a special point at $a = (1, 0)$. Now, we shall state some important rules for logarithms:

$$\textbf{(Log1)} \quad \ln(AB) = \ln A + \ln B$$

$$\textbf{(Log2)} \quad \ln\left(\frac{A}{B}\right) = \ln A - \ln B$$

$$\textbf{(Log3)} \quad \ln(A^k) = k \ln A$$

where $A, B > 0$ and k is a constant. Although we stated **Log1–Log3** for natural logarithms, the rules in question hold for logarithms to any base.

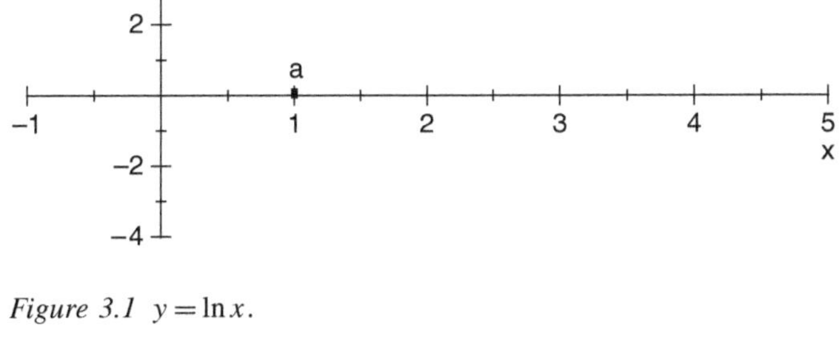

Figure 3.1 $y = \ln x$.

To prove **Log1** observe that, by definition of a logarithm,

$$e^{\ln(AB)} = AB$$
$$e^{\ln A} = A$$
$$e^{\ln B} = B$$

Therefore,

$$e^{\ln(AB)} = AB = e^{\ln A} e^{\ln B} = e^{\ln A + \ln B}$$

It follows rather readily that

$$\ln(AB) = \ln A + \ln B$$

To prove **Log2** observe that

$$\ln\left(\frac{A}{B}\right) = \ln(AB^{-1}) = \ln A + \ln B^{-1}$$
$$= \ln A + (-1)\ln B = \ln A - \ln B$$

Now, let us prove **Log3**. Note that

$$\ln A^k = \ln(A * A * A * \cdots * A)$$
$$= \underbrace{\ln A + \ln A + \cdots + \ln A}_{k \text{ times}} = k \ln A$$

EXAMPLE 3.2 Express the following logarithms in terms of $\ln 3$.

(a) $\ln 243$
(b) $\ln(\frac{1}{27})$
(c) $\ln(\sqrt[3]{27^2})$
(d) $\ln 3e$

Solution:

(a) $\ln 243 = 5 \ln 3$
(b) $\ln(\frac{1}{27}) = \ln 3^{-3} = -3 \ln 3$
(c) $\ln(\sqrt[3]{27^2}) = \ln(\sqrt[3]{3^6}) = \ln(3^{\frac{6}{3}}) = \ln 3^2 = 2 \ln 3$
(d) $\ln(3e) = \ln 3 + \ln e = 1 + \ln 3$

EXAMPLE 3.3 Solve the following equations for w.

(a) $17e^{10w+2} = 23$
(b) $2.7e^{\alpha w} = K$
(c) $e^w + \alpha e^w + \alpha^2 e^w = 17$

Solution:

(a) Divide each side of the equation by 17 to obtain

$$e^{10w+2} = \frac{23}{17}$$

Then take the ln of each side of the equation above to obtain

$$\begin{aligned}
\ln e^{10w+2} &= \ln\left(\frac{23}{17}\right) \\
10w + 2 &= \ln\left(\frac{23}{17}\right) \\
w &= \frac{\ln\left(\frac{23}{17}\right) - 2}{10}
\end{aligned}$$

(b) Divide each side of the equation by 2.7 and then take the natural logarithm of each side to obtain

$$\begin{aligned}
\alpha w &= \ln\left(\frac{K}{2.7}\right) \\
w &= \frac{1}{\alpha}\ln\left(\frac{K}{2.7}\right)
\end{aligned}$$

(c) First factor out e^w on the left-hand side to obtain

$$e^w(1+\alpha+\alpha^2) = 17$$

Then divide each side of the equation above by $(1+\alpha+\alpha^2)$ and take the ln of both sides to get

$$\ln\left(e^w\right) = \ln\left(\frac{17}{1+\alpha+\alpha^2}\right)$$

or

$$w = \ln\left(\frac{17}{1+\alpha+\alpha^2}\right)$$

3.2 Exponential functions

Suppose you have just opened a savings account that earns 5 percent interest a year and deposited \$1000 in the account. The amount of money you will have in your account after t years is given by

$$A(t) = 1000(1+0.05)^t \tag{3.1}$$

Using equation (3.1) we can find that

$$A(0) = 1000.00$$
$$A(1) = 1050.00$$
$$A(2) = 1102.5$$
$$A(3) = 1157.63$$
$$A(4) = 1215.50$$

So, each year your savings account grows by a factor of 1.05. Indeed,

$$A(1) = A(0) \cdot 1.05$$
$$A(2) = A(1) \cdot 1.05$$
$$A(3) = A(2) \cdot 1.05$$

and so on. A quantity that increases by a fixed factor per some unit of time (in our case per year) is said to *increase exponentially.* Thus, the amount of money in the savings account increases exponentially and we can think of the 5 percent interest rate on the account as the growth rate of the savings account. How long will it take for the account to double, triple or quadruple? In other words, what is the doubling, tripling or quadrupling time of the account?

To find the doubling time of the account, we solve the equation

$$1000(1 + 0.05)^{t_2} = 2000 \tag{3.2}$$

for t_2 to find that

$$t_2 = 14.21$$

To find the tripling time of the account, we solve

$$1000(1 + 0.05)^{t_3} = 3000 \tag{3.3}$$

for t_3 to find that

$$t_3 = \frac{\ln 3}{\ln(1.05)} = 22.52$$

To find the quadrupling time of the account, we solve

$$1000(1 + 0.05)^{t_4} = 4000 \tag{3.4}$$

for t_4 to find that

$$t_4 = \frac{\ln 4}{\ln(1.05)} = 28.41$$

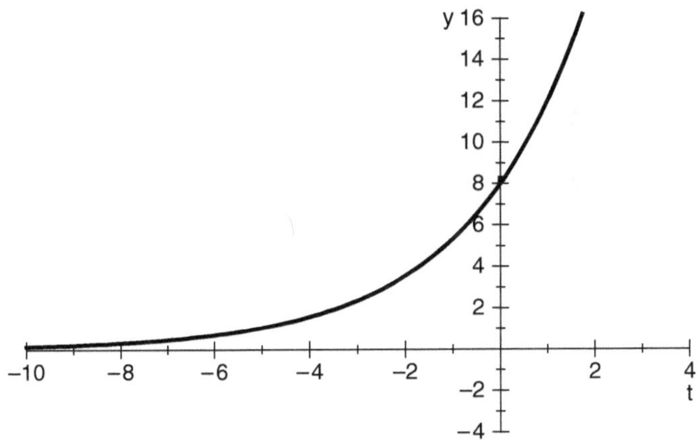

Figure 3.2 $y(t) = 8 \cdot 1.5^t$.

Now, the general exponential function with base a is given by

$$f(t) = Pa^t, \quad \text{where } a > 0 \tag{3.5}$$

and t is usually referred to as time.

The function in (1) is an exponential function with $P = 1000$ and $a = 1.05$. Exponential functions play a very important role in many economic, financial and demographic models. Consider, as another example, the graph of an exponential function with $P = 8$ and $a = 1.5$ as shown in Figure 3.2.

EXAMPLE 3.4 At the rate of inflation of 8 percent per year, how long will it take for the price level to double?

Solution: Let P denote the current price level. Doubling the price means

$$P(1 + 0.08)^{t_2} = 2P$$

Dividing the equation above by P and taking the ln of both sides yields

$$\ln(1 + 0.08)^{t_2} = \ln 2$$

Therefore

$$t_2 = \frac{\ln 2}{\ln(1 + 0.08)} \approx 9.0 \text{ (years)}$$

EXAMPLE 3.5 The population of Bolivia in 1950 was estimated to be three million people. In 1989, Bolivia's population was estimated to be seven million people. What was the average rate of growth g of Bolivia's population from 1950 to 1989?

Solution: Assuming the exponential growth model, we expect the population to depend on time t in years as

$$3{,}000{,}000(1+g)^t$$

Setting up an equation for g we have

$$3{,}000{,}000(1+g)^{49} = 7{,}000{,}000$$

and thus

$$(1+g)^{49} = \frac{7}{3}$$

$$[(1+g)^{49}]^{\frac{1}{49}} = \left(\frac{7}{3}\right)^{\frac{1}{49}}$$

$$(1+g) = \frac{7}{3}^{\frac{1}{49}}$$

$$g \approx 0.017$$

So, the average annual rate of growth g for Bolivia's population from 1950 to 1989 was around 1.7 percent.

EXAMPLE 3.6 During the 1990s the annual rate of inflation in the US was around 3 percent. Assuming that the rate of inflation in the US will stay at 3 percent per year, we can predict the price of an item that initially costs P_0 after t years using the formula

$$P(t) = P_0(1.03)^t$$

So, what is the predicted price of a \$200,000 house after 25 years?

Solution: The predicted price of a \$200,000 house after 25 years is

$$P(25) = 200{,}000(1+0.03)^{25} \approx \$418{,}755.59$$

3.3 Mathematica examples

We can plot the natural logarithm function using the following Mathematica code:

```
> Plot[{Log[x]}, {x, 0, 20}]
```

The code and the graph in Figure 3.3 suggest that the function was defined on the following domain:

$$\text{Domain} = \{x \in \mathbb{R} \mid 0 < x \leq 20\}$$

Mathematica allows us to plot two or more functions in one graph:

```
> Plot[{Log[x], Exp[x]}, {x, -2, 2}]
```

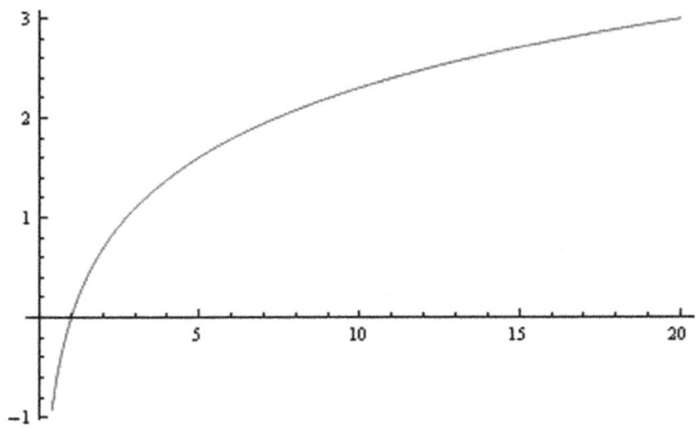

Figure 3.3

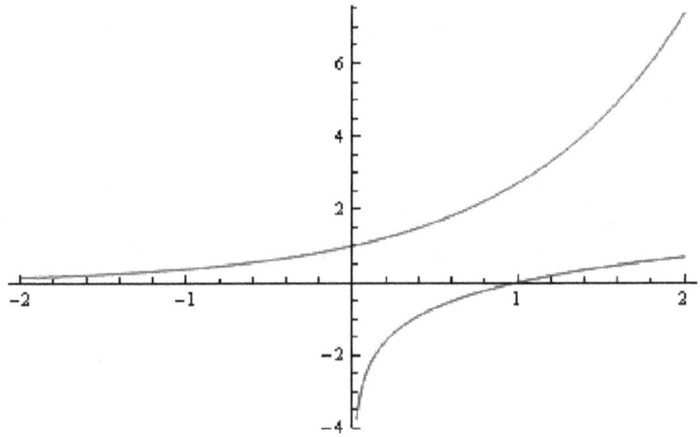

Figure 3.4

The code above plots both exponential and logarithmic functions in one graph, as shown in Figure 3.4.

PROBLEMS FOR SECTION 3.3

1. Express the following in terms of $\ln 5$.

   (a) $\ln 625$
   (b) $\ln \sqrt[3]{25}$
   (c) $\ln \frac{1}{125}$
   (d) $\ln \sqrt[2]{0.2}$

2. Solve the following equations for w.

   (a) $\ln w = e + 2$
   (b) $\ln(2w + 5) = 3$
   (c) $\ln(w^3 - w^2) = 0$
   (d) $\dfrac{w^3 \ln(w+3)}{(w^2 + 2w)} = 0$

3. Which of the following statements are true?

   (a) $\ln(ABCD) = \ln A + \ln B + \ln C + \ln D$
   (b) $\ln\left(\dfrac{ABC}{DE}\right) = \ln A + \ln B + \ln C - \ln D - \ln E$
   (c) $\ln\left(\dfrac{A}{B}\right) = 1$ implies that $A \neq B$
   (d) $B \ln A^B = B + B \ln A$
   (e) $\ln(A^{B+C}) = \ln A^B + \ln A^C$
   (f) $\ln A + \ln B - \ln C = \ln(AB - C)$

4. In 2002 the gross domestic product (GDP) of Austria was estimated to be $2.1 \cdot 10^{11}$ US dollars, and the rate of growth of GDP was estimated to be $r = 0.07$. If the GDP of Austria continued to grow exponentially at the rate $r = 0.07$, when would the GDP of Austria

   (a) triple?
   (b) quadruple?

5. In 2002 the GDP of Austria was estimated to be $2.1 \cdot 10^{11}$ US dollars, and the rate of growth of GDP was estimated to be $r = 0.07$. If the GDP of Austria continued to grow exponentially at the rate $r = 0.07$, when would the GDP of Austria double?

4 Limits and derivatives

In mathematics, differential calculus is a subfield of calculus that is concerned with the study of how quickly functions change over time. The primary concept in differential calculus is the derivative function. The derivative allows us to find the rate of change of economic variables over time.

This chapter introduces the concept of a derivative and lays out the most important rules of differentiation. To properly introduce derivatives, one needs to consider the idea of a limit. We cover the concept of a limit in the first section. The chapter closes with growth rates of discrete and continuous variables.

4.1 Limits

Consider a function g given by

$$g(x) = \frac{e^{2x} - 1}{x}$$

and shown in Figure 4.1. Clearly, the function is undefined for $x = 0$, since anything divided by zero is undefined. However, we can still ask what happens to $g(x)$ when x is slightly above or below zero. Using a calculator we can find the values of $g(x)$ in the neighborhood of $x = 0$, as shown in Table 4.1.

As x approaches zero, $g(x)$ takes values closer and closer to 2. So we can say that $g(x)$ tends to 2 as x tends to zero. We write

$$\lim_{x \to 0} \frac{e^{2x} - 1}{x} = 2$$

and say that the limit of $g(x)$ as x approaches zero is equal to 2.

Now that the idea of a limit is clear on an intuitive level, let us consider a formal definition of the right- and left-hand side limits.

Let f be a function defined on some open interval (a, b). We say that L is the right-hand side limit of $f(x)$ as x approaches a from the right and write

$$L = \lim_{x \to a^+} f(x)$$

if for every $\varepsilon > 0$ there is a $\delta > 0$ such that

$$|f(x) - L| < \varepsilon$$

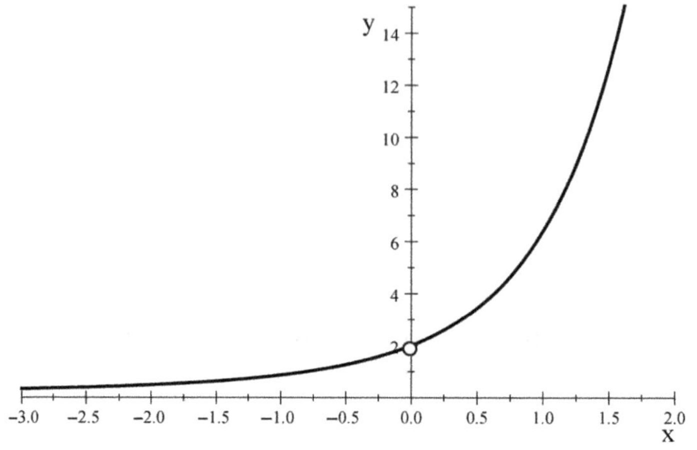

Figure 4.1

Table 4.1

x	−0.01	−0.001	−0.0001	0	0.0001	0.001	0.01
$g(x)$	1.980	1.998	1.9998	undef.	2.0002	2.002	2.020

whenever

$$a < x < a + \delta$$

As an example, let us consider the following function

$$f(x) = \begin{cases} x + 1, & x \geq 0 \\ x - 3, & x < 0 \end{cases}$$

We want to show that

$$\lim_{x \to 0+} f(x) = 1$$

Let us choose $\varepsilon > 0$. We need to show that there is a $\delta > 0$ such that

$$|f(x) - L| < \varepsilon$$

whenever

$$0 < x < \delta$$

Let us choose $\delta = (\varepsilon/2)$. Then,

$$f(x) - L = x + 1 - 1 = x$$

and therefore

$$0 < f(x) - L < \frac{\varepsilon}{2} < \varepsilon$$

It follows immediately that

$$|f(x) - L| < \varepsilon$$

whenever

$$0 < x < \delta$$

Now we have proved that the limit of

$$f(x) = \begin{cases} x+1, & x \geq 0 \\ x-3, & x < 0 \end{cases}$$

as x approaches zero from the right is equal to 1.

Now let us define a left-hand side limit. Let f be a function defined on some open interval (a, b). We say that L is the left-hand limit of $f(x)$ as x approaches b and write

$$L = \lim_{x \to b^-} f(x)$$

if for every $\varepsilon > 0$ there is a $\delta > 0$ such that

$$|f(x) - L| < \varepsilon$$

whenever

$$b - \delta < x < b$$

It could be shown easily that for the function above

$$\lim_{x \to 0^-} f(x) = -3$$

We say that a two-sided limit (or simply a limit) of $f(x)$ exists at point a when the left-hand side limit is equal to the right-hand side limit. For example, $f(x)$ above does not have a limit at $x = 0$ since

$$\lim_{x \to 0^-} f(x) = -3 \neq 1 = \lim_{x \to 0+} f(x)$$

Indeed, $f(x)$ has a jump at $x = 0$, so it is discontinuous. Intuitively we can call a function discontinuous at $x = a$ if it has a jump at that point. Now, let us formally define a continuous function.

A function f is continuous at $x=a$ if

(1) the two-sided limit L exists at $x=a$,

$$\lim_{x\to a} f(x)=L$$

(2) the value of f at $x=a$ equals L,

$$f(a)=L$$

For instance, the function $f(x)=x$ is continuous at every point.

We say that f is continuous if it is continuous at every point of its domain. In fact, all polynomial function are continuous.

4.1.1 *Infinite limits and limits at infinity*

Let f be a function defined on the interval

$$(b,a)\cup(a,c)$$

We say that the two-sided limit of f at $x=a$ equals ∞ and write

$$\lim_{x\to a} f(x)=\infty$$

if for every $N>0$ there is a $\delta>0$ such that

$$f(x)>N$$

whenever

$$0<|x-a|<\delta$$

EXAMPLE 4.1 Consider the function

$$f(x)=\frac{1}{x^2}$$

depicted in Figure 4.2. It is clear that

$$\lim_{x\to 0}\frac{1}{x^2}=\infty$$

4.2 First- and second-order derivatives

Consider a function $h(x)$ as depicted in Figure 4.3.

Two points A and B belong to the graph of the function, where A is fixed. A line that passes through points A and B on the graph of the function is called a *secant line*. Now, pick a point B_1 that belongs to the plot of $h(x)$, such that B_1 is in between A and B, and draw a

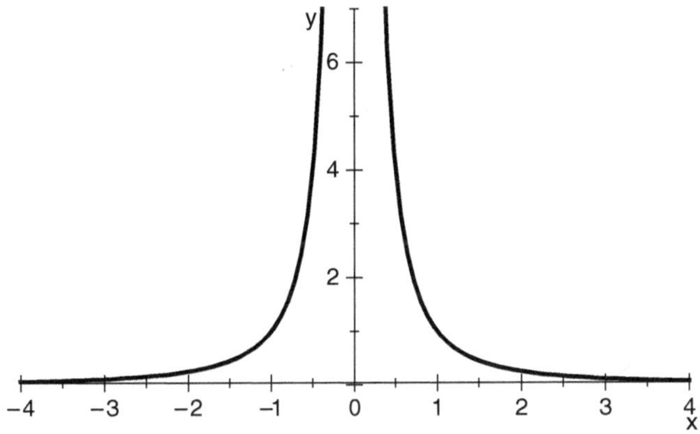

Figure 4.2 $y = \frac{1}{x^2}$.

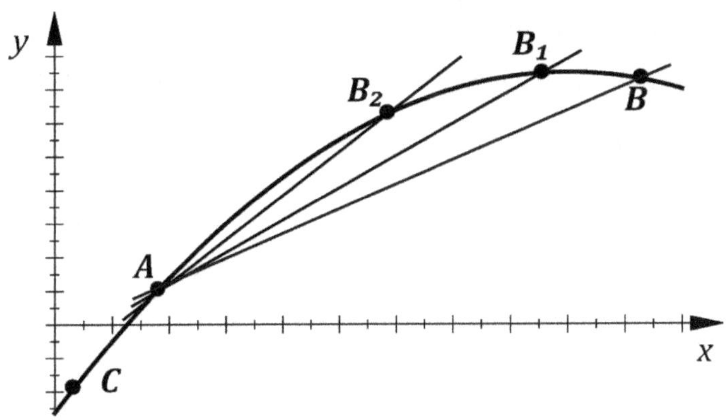

Figure 4.3 $y = h(x)$.

secant line through A and B_1. As point B is brought toward point A, the secant line tends to the right tangent line. The slope of the right tangent line is the limit of the slopes of the secant lines. The left tangent line to the curve $f(x)$ at point A is defined similarly: we start with the secant line AC, and as point C is brought toward A, the secant line tends toward the left tangent line. Clearly, the left and the right tangent lines pass through the same point A. If the slope of the left tangent line equals the slope of the right tangent line, then we say that $h(x)$ has a tangent line at point A. The tangent line at point A is the same line as the right (or left) tangent line at point A.

There are functions that do not have a tangent line at some or even every point.

For example, consider the function

$$f(x) = |x| = \begin{cases} x & \text{if } x \geq 0 \\ -x & \text{if } x < 0 \end{cases}$$

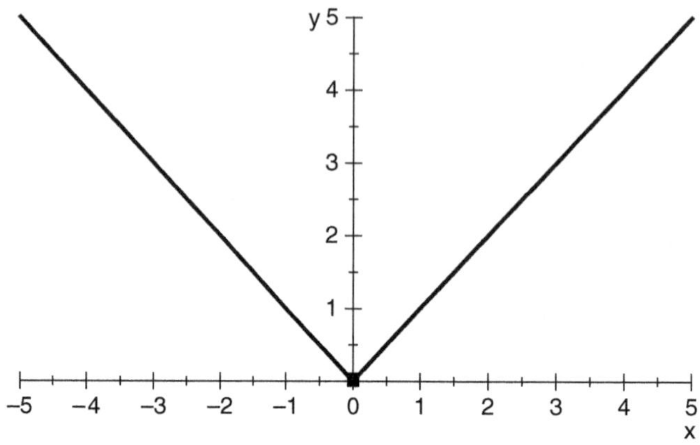

Figure 4.4 $y = |x|$.

depicted in Figure 4.4. The right tangent line is not equal to the left tangent line at point $P = (0, 0)$. Indeed, at point $P = (0, 0)$ the slope of the right tangent line is equal to 1, and the slope of the left tangent line is equal to -1.

DEFINITION 4.1 The *derivative* of a function $y = f(x)$ is the function $f'(x)$ defined by

$$f'(x) = \frac{dy}{dx} = \lim_{h \to 0} \frac{f(x+h) - f(x)}{h}$$

for all x for which the limit exists.

For instance, consider the function

$$y = f(x) = x^2$$

Then

$$\begin{aligned} f'(x) &= \frac{dy}{dx} \\ &= \lim_{h \to 0} \frac{(x+h)^2 - x^2}{h} \\ &= \lim_{h \to 0} \frac{x^2 + 2xh + h^2 - x^2}{h} \\ &= \lim_{h \to 0} \left(\frac{2hx + h^2}{h} \right) \\ &= \lim_{h \to 0} (2x + h) = 2x \\ f'(x) &= \frac{dy}{dx} = 2x \end{aligned}$$

The derivative of $f(x) = x^2$ is the function $f'(x) = 2x$.

EXAMPLE 4.2 Apply the definition of the derivative to differentiate $f(x)=x^3$.

Solution:

$$\begin{aligned} f'(x) &= \lim_{h\to 0}\frac{(x+h)^3-x^3}{h} \\ &= \lim_{h\to 0}\left(\frac{x^3+3x^2h+3xh^2+h^3-x^3}{h}\right) \\ &= \lim_{h\to 0}\left(\frac{3x^2h+3xh^2+h^3}{h}\right) \\ &= \lim_{h\to 0}(3x^2+3xh+h^2) \\ f'(x) &= 3x^2 \end{aligned}$$

Normally, we will not use the definition of the derivative to find the derivative of $f(x)$. We will use the general rules listed below. Let $u=u(x)$ and $v=v(x)$ and let k be a constant. Then the following rules hold:

(Der1) $(u^k)'=ku^{k-1}$

(Der2) $(ku)'=ku'$

(Der3) $(u\pm v)'=u'\pm v'$

(Der4) $(uv)'=u'v+uv'$

(Der5) $\left(\frac{u}{v}\right)'=\frac{u'v-uv'}{v^2}$

(Der6) $(k)'=0$

(Der7) $(\ln x)'=\frac{1}{x}$

(Der8) $(e^x)'=e^x$

Note, that **Der3** could be extended to any number of functions, as follows:

$$(u_1\pm u_2\pm\cdots\pm u_n)'=u_1'\pm u_2'\pm\cdots\pm u_n'$$

The same is true of **Der4**:

$$(v_1v_2v_3\ldots v_n)'=v_1'v_2v_3\ldots v_n+v_1v_2'v_3\ldots v_n+\cdots+v_1v_2v_3\ldots v_n'.$$

EXAMPLE 4.3 We use **Der1–Der8** to compute the derivatives of the following functions:

(a) $(x^5+2x^4+x^3-x^2+17)'=5x^4+8x^3+3x^2-2x$
(b) $[(x^2+3x)(3x^3-2x)]'=(2x+3)(3x^3-2x)+(x^2+3x)(9x^2-2)$
(c) $(27)'=0$

(d) $\left(\frac{x^2+2}{x^3+5}\right)' = \frac{(2x)(x^3+5)-(x^2+2)(3x^2)}{(x^3+5)^2}$

(e) $(5x^{\frac{4}{3}} - 3x^{-7})' = 5 \times \frac{4}{3}x^{\frac{1}{3}} + 21x^{-8}$

Observe that derivatives themselves are functions of the independent variable x. Let us, for example, consider $f(x)=x^2+12$ and its derivative function $f'(x)=2x$. Clearly, a different value of x will result in a different value of the derivative, such as

$$f'(1)=2$$
$$f'(5)=10$$
$$f'(-3)=-6$$

Observe that the slope of the tangent line to the graph of $f(x)=x^2+12$ at point $x=1$ equals 2. In general, we should remember that

$$f'(p) = \text{the slope of the tangent line to the graph of } f(x) \text{ at } p$$

The derivative of a function $f(x)$ is normally referred to as the *first derivative* of $f(x)$. Economists usually refer to $f(x)$ as the total function and $f'(x)$ as the marginal function. Often $f'(x)$ is also differentiable and in that case $f''(x)$ denotes the second derivative of $f(x)$. There are three different forms of notation for the second derivative of a function $y=f(x)$ at a given point x:

$$f''(x) = \frac{d^2y}{dx^2} = y''$$

EXAMPLE 4.4 Let $f(x)=x^4+2x^2+17$. Find $f'(x)$ and $f''(x)$.

Solution: Using **Der1–Der8** we first differentiate the total function

$$f'(x)=4x^3+4x$$

and then differentiate the marginal function to obtain

$$f''(x)=12x^2+4$$

Of course, a function $y=f(x)$ may be differentiable more than two times. Sometimes we may be interested in finding the third, fourth or nth derivative of a function. If $y=f(x)$ is differentiable n times, then we let $f^{(n)}(x)$ denote the nth derivative of $y=f(x)$ at a given point x.

EXAMPLE 4.5 Let $w(x)=x^7+2x^5+19$. Find the first four derivatives of $w(x)$.

Solution: Using **Der1–Der8**, we have

$$w'(x) = 7x^6 + 10x^4$$
$$w''(x) = 42x^5 + 40x^3$$
$$w^{(3)}(x) = 210x^4 + 120x^2$$
$$w^{(4)}(x) = 840x^3 + 240x$$

Note that $w(x) = x^7 + 2x^5 + 19$ is a polynomial function and as such can be differentiated infinitely many times. However, all derivatives of order eight or above are equal to zero. In general, if $f(x)$ is a polynomial function of degree n, then the first n derivatives of $f(x)$ are non-zero and all derivatives of higher order vanish.

EXAMPLE 4.6 Consider a firm producing output with just one input, labor L. The firm's total product Q is a function of its only input: $Q(L) = AL^{1-b}$. Find $Q'(L)$ and $Q''(L)$.

Solution: Using **Der1–Der8**, we have

$$Q'(L) = A(1-b)L^{-b}$$
$$Q''(L) = A(1-b)(-b)L^{-b-1}$$

Economists refer to the expression

$$\frac{dQ}{dL} = Q'(L)$$

as the marginal product of labor (MPL). The MPL is defined as the change in total product from expanding labor input by one unit while holding everything else constant. By assumption MPL is always positive.

Functions that are continuous and have continuous first-order derivative belong to a class of functions called C^1. This notation is commonly used in mathematics and economics books. The index 1 in the notation C^1 indicates that all the derivatives up to the first order exist and are continuous. Similarly, notation C^2 stands for functions that are twice differentiable and their second derivatives are continuous.

4.2.1 Partial derivatives

Often economists work with multivariate functions such as

$$f(x, y, w) = x^2 + y^3 + w^5$$

We can differentiate the function above with respect to each of its three independent variables as follows:

$$\frac{\partial f}{\partial x} = f_1(x, y, w) = 2x$$

$$\frac{\partial f}{\partial y} = f_2(x, y, w) = 3y^2$$

$$\frac{\partial f}{\partial w} = f_3(x, y, w) = 5w^4$$

Note that whenever we differentiate a multivariate function with respect to one independent variable, we consider the remaining variables to be constant. The symbol for the partial derivative, ∂, is used instead of the normal derivative symbol, d, in a situation when a function depends on several variables and we differentiate with respect to one of those variables. Subscript 1 in expression $f_1(x, y, w)$ denotes differentiation with respect to the function's first variable, which is x in our case.

The same notation applies to the second- and higher-order derivatives. For instance, for function $g(x, y) = 3x^2y^4$, the symbol $g_{12}(x, y)$ would mean

$$g_{12}(x, y) = \frac{\partial^2 g}{\partial y\, \partial x} = 24xy^3$$

4.3 The chain rule

Suppose that w is a differentiable function of u, and u is in turn a differentiable function of x, then

$$\frac{dw}{dx} = \frac{dw}{du}\frac{du}{dx} \tag{4.1}$$

The statement above is known as the *chain rule* and it has numerous applications in both mathematics and economics. Since $w(x)$ is a composite function, the chain rule is also referred to as the *composite-function rule*. Some functions, such as $w(x) = (3x^5 + 7x^2)^{25}$, would be very difficult to differentiate without the use of the chain rule. The extension of the chain rule to the case of three functions is straightforward. If w is a differentiable function of u, and u is a differentiable function of x, and x is a differentiable function of t, then

$$\frac{dw}{dt} = \frac{dw}{du}\frac{du}{dx}\frac{dx}{dt} \tag{4.2}$$

The chain rule can be easily extended to the case of $n \geq 4$ functions in a similar fashion.

EXAMPLE 4.7 Using the chain rule find $\frac{dy}{dx}$ when $y(u) = u^4$ and $u(x) = 5 - 3x^2$.

Solution: Since $\frac{dy}{du} = 4u^3$ and $\frac{du}{dx} = -6x$, we have

$$\frac{dy}{dx} = \frac{dy}{du}\frac{du}{dx} = 4u^3(-6x) = 4(5 - 3x^2)^3(-6x) = -24x(5 - 3x^2)^3$$

EXAMPLE 4.8 Using the chain rule find $\frac{dy}{dx}$ when $y(u) = u^c$ and $u(x) = ax + b$.

Solution: Observe that $\frac{dy}{du} = cu^{c-1}$ and $\frac{du}{dx} = a$. Thus

$$\frac{dy}{dx} = \frac{dy}{du}\frac{du}{dx} = cu^{c-1}a = ac(ax+b)^{c-1}$$

EXAMPLE 4.9 Using the chain rule find $\frac{dy}{dx}$ when $y = \sqrt[3]{u}$ and $u = 3x^2 + 2x + 17$.

Solution: Note that $\frac{dy}{du} = \frac{1}{3}u^{-\frac{2}{3}}$ and $\frac{du}{dx} = 6x + 2$. So

$$\frac{dy}{dx} = \frac{dy}{du}\frac{du}{dx} = (\sqrt[3]{3x^2+2x+17})' = \frac{1}{3}(6x+2)(3x^2+2x+17)^{-\frac{2}{3}}$$

Recall that derivative represents rate of change. In a situation when we have quantity y depending on another quantity u, which in turn depends on another quantity x, we expect the total rate of change, y with respect to x, to be the product of the rates. That is the gist of the chain rule.

PROBLEMS FOR SECTION 4.3

1. Find the first derivative of the following functions.

(a) $y(x) = 5x^4 - 4x^3 + 3x^2 - 21x + 15$
(b) $g(x) = 5e^x + \ln x - 2x^{0.5}$
(c) $h(x) = \frac{ax+b}{cx}$
(d) $k(x) = (x^3+1)(x^2+1)(x+1)a$
(e) $w(x) = ax^4 + b(x+1)(cx^2+2)$
(f) $u(x) = \frac{x+4}{x^3+2}$
(g) $f(x) = (e^x+1)(\ln x+2)(x^4+3)$
(h) $r(x) = \frac{(\ln x+1)}{(e^x+2)}$
(i) $t(x) = (x^2+3)(x^3-4)(x^4+5)(x^5-6)(x^6+7)$

2. Find the first and second derivatives of the following functions.

(a) $y(x) = 14x^3 + 3x^2 - 21x + 23$
(b) $g(x) = \frac{ax+b}{cx}$
(c) $h(x) = (ax^2 - b)(cx^5 + dx)$
(d) $f(x) = \frac{1}{x^3}$
(e) $t(x) = x^{-5}(x^4+1)\sqrt[5]{x}$
(f) $z(x) = \frac{1}{\sqrt[5]{x^3}}$

(g) $p(x) = \frac{2x+1}{3x-7}$

(h) $v(x) = \frac{5x+1}{x^7}$

(i) $d(x) = \frac{5x-9}{x+4}$

(j) $w(x) = 5x^{-5} + x^{-4}$

(k) $k(x) = \frac{4x-5}{3x^2+x+1}$

3. Use the chain rule to find $\frac{dy}{dx}$ for the following functions.

(a) $y = \frac{20}{(x^4+4x+5)^8}$

(b) $y = (x^5+x^2)^{99}$

(c) $y = \left(\frac{x-1}{x+6}\right)^{\frac{1}{3}}$

(d) $y = \sqrt{x^5+17}$

(e) $y = (25x+x^{25})^3$

(f) $y = \left(\frac{x+2}{100x^4+2x+10}\right)^4$

(g) $y = \left(\frac{x+2}{x-1}\right)^{\frac{1}{2}}$

(h) $y = 5 + (x^5+x^4+x^3+x^2+x+1)^6$

(i) $y = \sqrt[4]{(x^4+2x^2+12)^3}$

(j) $y = \frac{1}{3}(2x^4+5)^{88}$

4. Use the chain rule to find $\frac{dy}{dx}$ for the following functions.

(a) $y = 3\sqrt[7]{1+\sqrt[5]{2+x^3}}$

(b) $y = 7[21+(3x^2+1)^{40}]^{85}$

(c) $y = [125+\sqrt[3]{(2x^6+17)^{22}}]^{15}$

(d) $y = [72x+(x^{16}-2x^3)^6]^{\frac{1}{2}}$

(e) $y = 1+[1+(x^2+1)^5]^5$

5. Let $W = h(k)$ and $k = f(t)$. Find the formula for $\frac{dW}{dt}$.

6. Let $F = g(v)$ and $v = f(t)$. Find the formula for $\frac{dF}{dt}$.

7. Suppose that the total cost function is $\text{TC} = 400 - 3(525-3Q)^{\frac{1}{4}}$, where $Q \leq 70$. Use the chain rule to find $\frac{d(\text{TC})}{dQ}$.

4.4 Total and marginal functions

A *total cost function* TC(Q) assigns to each level of output Q the total cost of producing that level of output. Naturally, total cost functions are strictly increasing functions and

$$\text{Domain(TC)} = \{Q \mid Q \geq 0\}$$

The derivative MC(Q) of the total cost function

$$\text{MC} = \frac{d(\text{TC})}{dQ}$$

is known as the *marginal cost function*. The *average cost function* AC is given by

$$\text{AC} = \frac{\text{TC}}{Q}$$

and it measures the cost per unit produced.

EXAMPLE 4.10 Let Q be a firm's output level and let TC be the total cost of production, given by

$$\text{TC} = aQ^2 + bQ + c, \quad \text{where } a, b, c > 0$$

Find the average cost AC and the marginal cost MC.

Solution:

$$\text{AC} = \frac{\text{TC}}{Q} = aQ + b + \frac{C}{Q}$$

$$\text{MC} = \frac{d(\text{TC})}{dQ} = 2aQ + b$$

EXAMPLE 4.11 A firm has a total revenue function given by

$$\text{TR} = 12Q - 0.2Q^2, \quad \text{where } 0 \leq Q \leq 30$$

The TR function is depicted in Figure 4.5. Find the marginal and average revenue functions MR and AR. Compare the slopes of MR and AR.

Solution: It follows rather readily that

$$\text{MR} = \frac{d(\text{TR})}{dQ} = 12 - 0.4Q$$

$$\text{AR} = \frac{\text{TR}}{Q} = 12 - 0.2Q$$

Clearly, the slope of the AR function is greater than the slope of the MR function (see Figure 4.6).

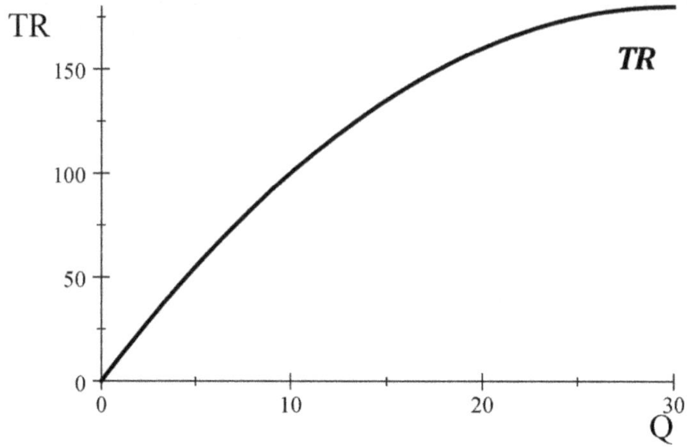

Figure 4.5 $\text{TR} = 12Q - 0.2Q^2$.

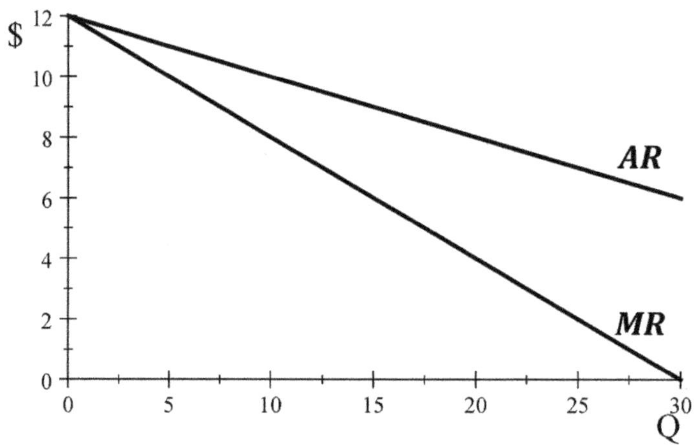

Figure 4.6

Again, we shall refer to a function $y = f(x)$ as the total function, and the derivative function $\frac{dy}{dx}$ shall be referred to as the marginal function. Thus, TR is typically the total function and MR is the marginal function.

Both total and marginal functions can be plotted on the same graph against the variable x. In Figure 4.7 a quadratic total function $g(x)$ is seen to have a linear marginal function $g'(x)$.

At point $x_0 = 6$ the marginal function takes a value of zero, which implies that at point $x_0 = 6$ the slope of the line tangent to the total function is equal to zero. In addition, at any point to the right (left) of x_0 the slope of the tangent line is positive (negative). For example,

$$f'(5) = -18 < 0$$

$$f'(6) = 0$$

$$f'(7) = 18 > 0$$

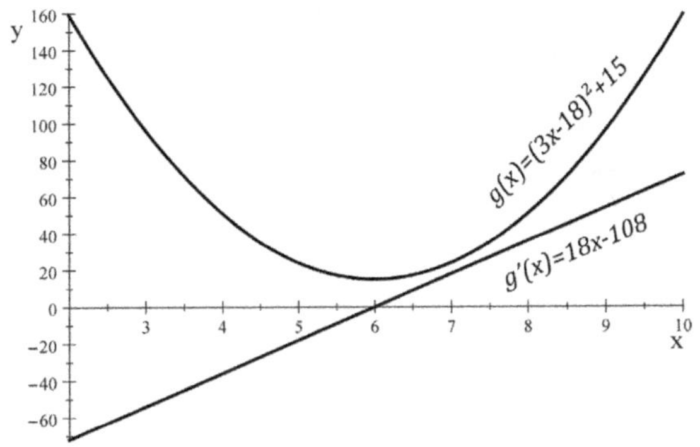

Figure 4.7

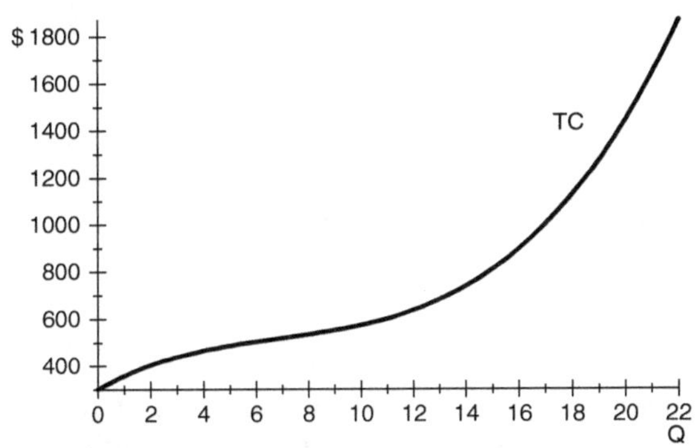

Figure 4.8 $\text{TC} = \frac{1}{3}Q^3 - 7Q^2 + 64Q + 300$.

EXAMPLE 4.12 Given the total cost function

$$\text{TC} = \frac{1}{3}Q^3 - 7Q^2 + 64Q + 300$$

shown in Figure 4.8, find the marginal and average cost functions.

Solution: It follows immediately that

$$\text{MC} = \frac{d(\text{TC})}{dQ} = Q^2 - 14Q + 64,$$

$$\begin{aligned}\text{AC} = \frac{\text{TC}}{Q} &= \frac{\frac{1}{3}Q^3 - 7Q^2 + 64Q + 300}{Q} \\ &= \frac{1}{3}Q^2 - 7Q + 64 + \frac{300}{Q}\end{aligned}$$

The graphs of MC and AC are shown in Figure 4.9.

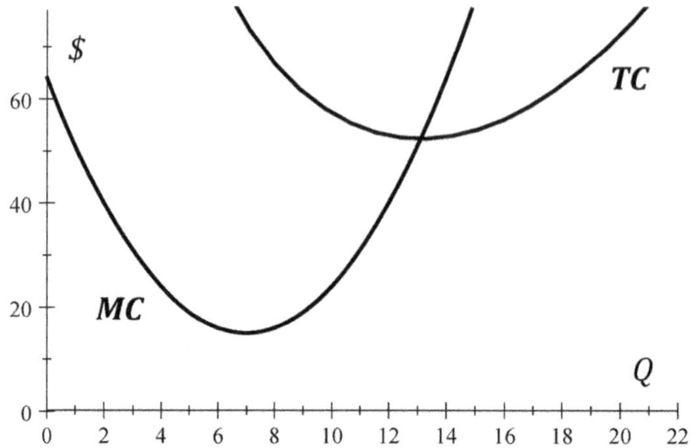

Figure 4.9

4.5 Growth rates

Consider an economic variable, say y, that changes over time. We write

$$y = f(t)$$

to indicate that our variable y is time-dependent. There are two methods of formalizing problems that involve time-dependent variables, known as discrete-time and continuous-time analysis. Economic variables are typically observed at regular time intervals (days, weeks, months) and as such are discrete.

Let us consider an economic variable $y = f(t)$, which changes discretely over time. In this case t takes values of $1, 2, 3, \ldots$ and y takes values y_1, y_2, y_3 and so on. The rate of change of the discrete variable y is

$$\textit{Rate of change of } y = y_{t+1} - y_t \tag{4.3}$$

The rate of growth of the discrete variable y is

$$\textit{Rate of growth of } y = \frac{y_{t+1} - y_t}{y_t} \tag{4.4}$$

EXAMPLE 4.13 Suppose that

$$y = at^2 + bt + c$$

where a, b, c are constants. Find the discrete-time rate of change and rate of growth.

Solution: It follows from (4.3) that

$$\begin{aligned} \textit{Rate of change of } y &= [a(t+1)^2 + b(t+1) + c] - [at^2 + bt + c] \\ &= 2at + a + b \end{aligned}$$

It follows from (4.4) that

$$\textit{Rate of growth of } y = \frac{2at + a + b}{at^2 + bt + c}$$

Often it is convenient to approximate discrete models with continuous ones. Let $y = f(t)$ be an economic variable that changes with time, where f is a differentiable function. Then

$$\textit{Rate of change of } y = \frac{dy}{dt} = f'(t) \tag{4.5}$$

and

$$\textit{Rate of growth of } y = \frac{f'(t)}{f(t)} = \frac{\dot{y}}{y} \tag{4.6}$$

EXAMPLE 4.14 Suppose that $y(t) = q + ct^2$, where q and c are constants. Find the rate of change of y and the rate of growth of y.

Solution: It follows from (4.5) that

$$\textit{Rate of change of } y = (q + ct^2)' = 2ct$$

and it follows from (4.6) that

$$\textit{Rate of growth of } y = \frac{2ct}{q + ct^2}$$

Discrete and continuous rates of change play a crucial role in micro- and macroeconomic analysis.

PROBLEMS FOR SECTION 4.5

1. The total cost of producing Q units of a good is

$$\text{TC}(Q) = \frac{1}{3}Q^3 - 7Q^2 + 64Q + 300$$

(a) Find the marginal cost function.
(b) Find the average cost function.
(c) Find the value Q for which the marginal cost is the least.
(d) Find the value of Q for which the average cost is the least.

2. Suppose that

$$y(t) = at + b$$

Find the continuous-time rate of growth.

3. Suppose that aggregate income Y and population size P are given by

$$Y(t) = a + bt, \quad P(t) = Ae^{rt}$$

where a, b, A and r are constants. Find the continuous-time growth rates of income, population and income per capita.

4. Suppose that $y(t) = A^{-rt}$, where A and r are constants. Find

   (a) the continuous-time rate of growth,
   (b) the discrete-time rate of growth.

5. Suppose that $y(t) = 5t^4$, where t is time. Find

   (a) the continuous-time rate of growth,
   (b) the discrete-time rate of growth.

6. Suppose that $y(t) = at^3 + bt^2 + ct + d$, where a, b, c and d are constants. Find

   (a) the continuous-time rate of growth,
   (b) the discrete-time rate of growth.

5 Optimization of univariate functions

Many problems in economics require identification of an optimal outcome. This task is achieved by maximizing or minimizing some economic entity. For example, a firm may seek to maximize profit π by choosing the optimal level of output Q. Since π is the difference between total revenue (TR) and total cost (TC) we can write

$$\pi(Q) = \mathrm{TR}(Q) - \mathrm{TC}(Q)$$

where $\pi(Q)$ is known as the objective function and Q is referred to as the choice variable. The objective function $\pi(Q)$ is the function that we seek to optimize. Other problems in economics require optimizing across two or more choice variables: one such example would be to minimize a cost function with respect to capital and labor. Another example would be to maximize a profit function with respect to capital, labor and land. In this chapter we will confine ourselves to examining objective functions with one variable, otherwise known in calculus as univariate functions. Finding extreme values of univariate functions is one of the most widely known and applied areas of calculus.

5.1 Local and global extrema

In mathematics, maxima and minima are known collectively as extrema. The following is a rigorous definition of local and global extrema.

DEFINITION 5.1 Let $f(x)$ be a function defined on an interval I and $A = (a, f(a))$ be a point on the graph of the function. Then $f(x)$ has:

(i) a *global maximum* at A if $f(a) \geq f(x)$ for all $x \in I$;
(ii) a *global minimum* at A if $f(a) \leq f(x)$ for all $x \in I$;
(iii) a *local maximum* at A if $f(a) \geq f(x)$ for all x in a neighborhood of a;
(iv) a *local minimum* at A if $f(a) \leq f(x)$ for all x in a neighborhood of a;
(v) a *strict global maximum* at A if $f(a) > f(x)$ for all $x \in I$;
(vi) a *strict global minimum* at A if $f(a) < f(x)$ for all $x \in I$;
(vii) a *strict local maximum* at A if $f(a) > f(x)$ for all x in a neighborhood of a;
(viii) a *strict local minimum* at A if $f(a) < f(x)$ for all x in a neighborhood of a.

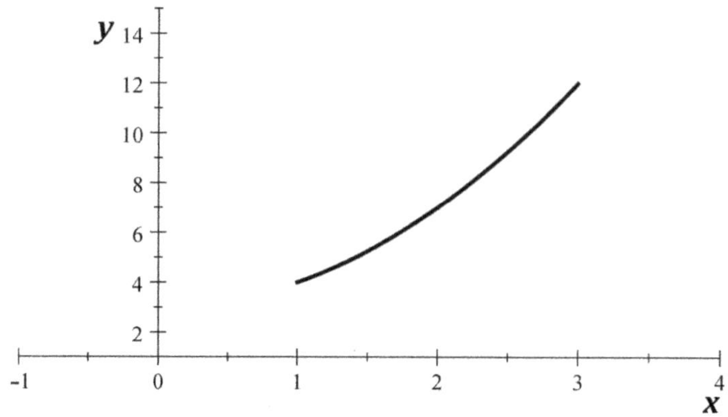

Figure 5.1

There are functions that do not have local or global extrema. For example, consider the function

$$g(x) = x^2 + 3, \quad \text{where } x \in I = (1, 3)$$

depicted in Figure 5.1.

Let us prove by contradiction that $g(x)$ does not have a global maximum. Suppose that $g(x)$ has a global maximum at some point $A = (a, f(a))$. Then, we have

$$g(a) = a^2 + 3$$

Since I is an open interval, there exists some $\varepsilon > 0$ such that $(a + \varepsilon) \in I$. This in turn implies that

$$g(a + \varepsilon) = (a + \varepsilon)^2 + 3 > g(a)$$

which is a contradiction.

There are many interesting cases as far as local and global extrema are concerned. We will discuss only two of these cases below.

Case 1 (a function has exactly one extremum). For example, the function depicted in Figure 5.2(a) has only one extremum – a (local and global) minimum at point A. And the function depicted in Figure 5.2(b) also happens to have only one extremum – a (local and global) maximum at point A. Note that both extrema in question are strict.

Case 2 (a function has several extrema). The function depicted in Figure 5.2(c) has three strict local (and global) minima at points A, C, E and two strict local maxima at points B and D. Further, a function may have more than one global maximum – the function in Figure 5.2(f) has three global maxima at points A, C and E. It is also possible that a function has only local extrema as shown in Figure 5.2(e). Indeed, the function has a local maximum at point A and a local minimum at point B, but has no global extrema.

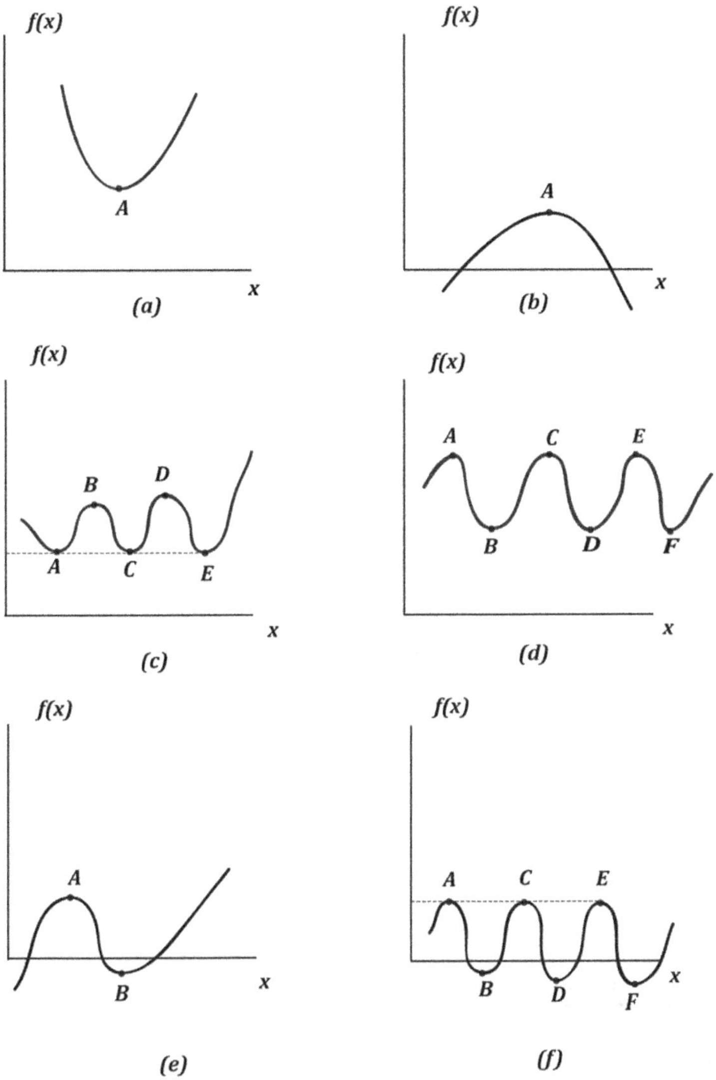

Figure 5.2

Most functions that economists encounter are differentiable at least once. Given that a function $f(x)$ is differentiable and for some point a we have

$$f'(a) = 0$$

we say that a is a *critical point* of $f(x)$ and point $A = (a, f(a))$ on the graph of the function is a *stationary point* of the function in question. Thus, to find stationary points of a function we must first find the critical points of the function and evaluate the function at the critical points.

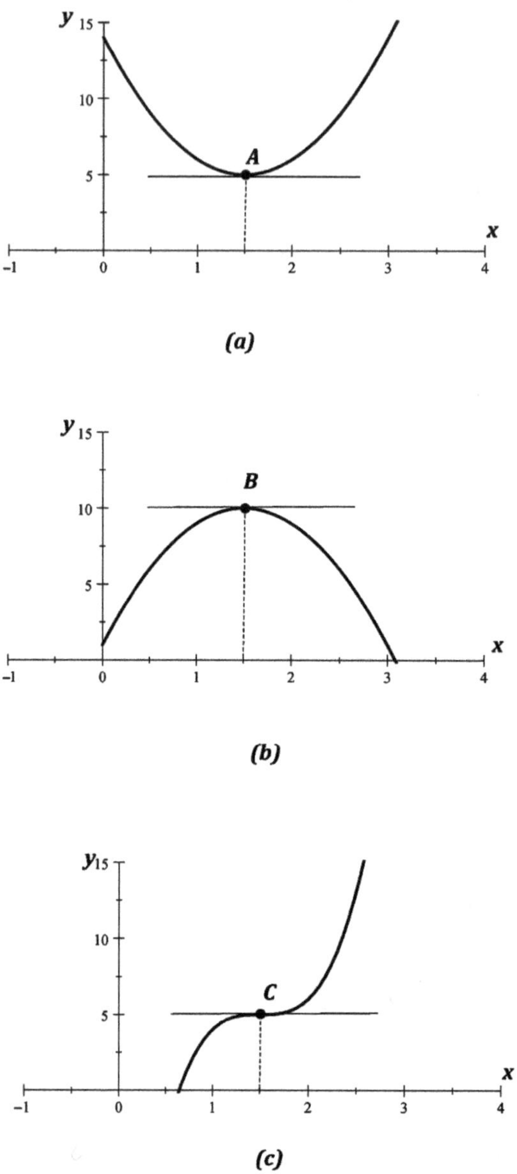

Figure 5.3

A stationary point could be a maximum, a minimum or an inflection point. Note that each function in Figure 5.3 has exactly one stationary point (A, B and C, respectively). Further, points A and B are local extrema while point C is an inflection point. Note that not all inflection points are stationary points.

In many important cases we can find the maximum or minimum values of a function just by studying the first derivative and its sign. Suppose $f(x)$ is differentiable in an open interval

I and that it has only one stationary point $A = (a, f(a))$. Further, let $f'(x) > 0$ for all x in I such that $x < a$, whereas $f'(x) < 0$ for all x in I such that $x > a$. Then $f(x)$ is increasing to the left of c and decreasing to the right of c. It follows that $f(a) > f(x)$ for all $x \neq c$. Hence, point A is a local maximum of $f(x)$.

With some trivial modifications, similar results hold for local minima.

5.1.1 Tests for local extrema

As you already know, if $f(x)$ has a local maximum or minimum at $(a, f(a))$, then a must be a critical point of $f(x)$. However, not every critical point gives rise to a maximum or minimum of a function (see Figure 5.3(c)). As you can see from Figure 5.2(b), the sign of $f'(x)$ changes from positive to negative at a, which tells us that $f(x)$ is increasing to the left of a and decreasing to the right of a. It follows rather readily that $f(x)$ has a local maximum at A. One of the best ways to classify a stationary point (i.e. to determine whether a stationary point is a local minimum, a local maximum or an inflection point) is to use the following tests for local extrema.

THEOREM 5.1: (FIRST-DERIVATIVE TEST FOR LOCAL EXTREMA). *Suppose that $f(x)$ is differentiable on an open interval I and it has only one stationary point $A = (a, f(a))$ on the interval.*

(a) *If $f'(x) > 0$ for $x < a$ and $f'(x) < 0$ for $x > a$, then A is a local maximum.*
(b) *If $f'(x) < 0$ for $x < a$ and $f'(x) > 0$ for $x > a$, then A is a local minimum.*
(c) *If $f'(x) < 0$ for $x < a$ and $x > a$, then A is an inflection point.*
(d) *If $f'(x) > 0$ for $x < a$ and $x > a$, then A is an inflection point.*

The theorem above requires knowing the sign of $f'(x)$ in the neighborhood of the stationary point in question. It is easy to remember the first-derivative test by visualizing diagrams such as those in Figure 5.3. There is yet another test for local extrema, namely the second-derivative test, that does not require knowing the sign of $f'(x)$ in the neighborhood of the stationary point. The second-derivative test below is probably the most popular test for local extrema.

THEOREM 5.2: (SECOND-DERIVATIVE TEST FOR LOCAL EXTREMA). *Suppose that $f(x)$ is twice differentiable on an open interval I and it has only one stationary point $A = (a, f(a))$ on that interval.*

(a) *If $f''(a) < 0$, then A is a local maximum.*
(b) *If $f''(a) > 0$, then A is a local minimum.*
(c) *If $f''(a) = 0$, then A may be a local extremum or an inflection point.*

As we can see from Theorem 5.2, sometimes the second derivative evaluated at a critical point may vanish. In such a case we can use the higher-order derivative test for local extrema to classify our stationary points.

THEOREM 5.3: (HIGHER-ORDER DERIVATIVE TEST FOR LOCAL EXTREMA). *Let f be a differentiable function on the interval I and let $a \in I$ be a critical point of $f(x)$ such that*

(i) $f^{(1)}(a)=f^{(2)}(a)=\cdots=f^{(n-1)}(a)=0$
(ii) $f^{(n)}(a)$ *exists and is non-zero.*

(a) *If n is even and $f^{(n)}(a)<0$, then $(a, f(a))$ is a local maximum.*
(b) *If n is even and $f^{(n)}(a)>0$, then $(a, f(a))$ is a local minimum.*
(c) *If n is odd, then $(a, f(a))$ is an inflection point.*

Note that the higher-order derivative test will work only if the function $f(x)$ yields a non-zero derivative value $f^{(n)}(a)$ for some finite number n. While most functions will meet this condition, there are some rare functions that produce a zero-valued derivative $f^{(n)}(a)$ in all successive differentiation.

EXAMPLE 5.1 Find the stationary points of the following functions:

(a) $y=2(x-3)^4+3$
(b) $y=x^5-7$

Use the higher-order derivative test to determine whether they represent local extrema or inflection points.

Solution:

(a) Since the first derivative of $f(x)$ is

$$f'(x)=8(x-3)^3$$

we have that $a=3$ is the critical point of the function. By successive differentiation of $f(x)$, continued until we encounter a non-zero derivative value at $a=3$, we obtain

$$\begin{aligned} f^{(2)}(x)&=24(x-3)^2 \Rightarrow f^{(2)}(3)=0\\ f^{(3)}(x)&=48(x-3) \Rightarrow f^{(3)}(3)=0\\ f^{(4)}(x)&=48 \Rightarrow f^{(4)}(3)=48>0 \end{aligned}$$

Since 4 is an even number and since $f^{(4)}(3)>0$, we conclude that $f(x)$ has a local minimum at $(3, f(3))$.

(b) Since the first derivative of $f(x)$ is

$$f'(x)=5x^4$$

we have that $a=0$ is the stationary point of the function. By successive differentiation of $f(x)$, continued until we encounter a non-zero derivative value at $a=0$, we obtain

$$\begin{aligned} f^{(2)}(x)&=20x^3 \Rightarrow f^{(2)}(0)=0\\ f^{(3)}(x)&=60x^2 \Rightarrow f^{(3)}(0)=0\\ f^{(4)}(x)&=120x \Rightarrow f^{(4)}(0)=0\\ f^{(5)}(x)&=120 \Rightarrow f^{(5)}(0)>0 \end{aligned}$$

Since 5 is an odd number, we conclude that $f(x)$ has an inflection point at $A=(0,-7)$.

Note that we could use the first-derivative test to classify the stationary points of the two functions in question. However, the solution process in that case would be more complicated.

EXAMPLE 5.2 Find the relative extrema of the function

$$f(x) = \frac{1}{3}x^3 + x^2 - 15x + 72$$

Solution: First, we find the derivative of the function to be

$$f'(x) = x^2 + 2x - 15$$

It follows that $f'(x) = 0$ at $x_1 = 3$ and $x_2 = -5$. In other words, $A = (-5, \frac{391}{3})$ and $B = (3, 45)$ are the stationary points of $f(x)$.

In the neighborhood of $x_1 = 3$, we have $f'(x) < 0$ for $x < 3$, and $f'(x) > 0$ for $x > 3$. Thus, B is a relative minimum. Similarly, in the neighborhood of $x = -5$, we find $f'(x) > 0$ for $x < -5$, and $f'(x) < 0$ for $x > -5$, confirming that A is a relative maximum (see Figure 5.4).

EXAMPLE 5.3 Find the minimum of the average cost function:

$$\text{AC} = f(Q) = Q^2 - 10Q + 32$$

Solution: Note that the graph of $f(Q)$ is a parabola that opens up, since the coefficient at the quadratic term is positive. According to Section 1.2.2, our quadratic function with $a = 1$, $b = -10$, and $c = 32$ has a global minimum at a point

$$A = \left(\frac{-b}{2a}, f\left(\frac{-b}{2a}\right)\right)$$

Since the value of the function at $x = 5$ is $f(5) = 7$, we write that the average cost function has a minimum at a point $A = (5, 7)$.

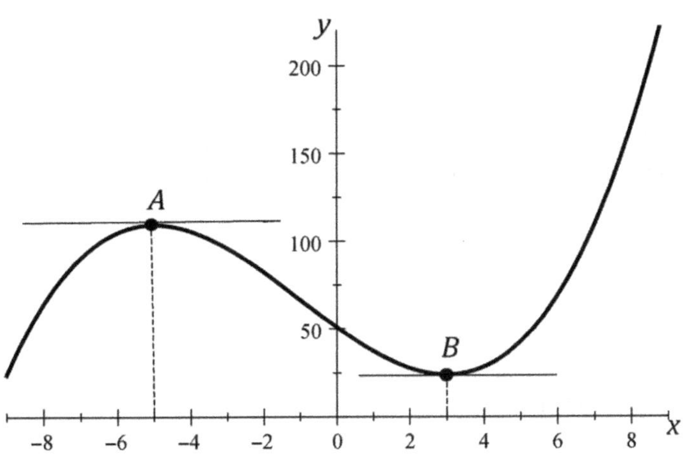

Figure 5.4

Example 5.4 Let the total revenue and total cost functions be

$$\text{TR}(Q) = 2.5Q - 2.25Q^2$$
$$\text{TC}(Q) = 27 - 80Q + 18Q^2 - Q^3$$

where TR and TC are in dollars and Q is in units (say, pounds per day). Find the profit-maximizing quantity of the output.

Solution: The profit function is

$$\begin{aligned}\pi = f(Q) &= \text{TR}(Q) - \text{TC}(Q)\\ &= 82.5Q - 20.25Q^2 + Q^3 - 27\end{aligned}$$

Observe that

$$\begin{aligned}\pi' = f'(Q) &= 82.5 - 40.5Q + 3Q^2\\ f'(Q) &= (3x - 7.5)(x - 11)\end{aligned}$$

This implies that the profit function has two critical points $x = 2.5$ and $x = 11$. Now, the second derivative is given by

$$f''(Q) = -40.5 + 6Q \quad \begin{cases} > 0 & \text{when } Q = 11 \\ < 0 & \text{when } Q = 2.5 \end{cases}$$

which implies that the profit-maximizing output is $Q^* = 2.5$ (pounds per day). Note that the other output level minimizes the profit. By substituting Q^* into the profit function, we can find the maximized profit π^* to be

$$\begin{aligned}\pi(Q) &= 82.5Q - 20.25Q^2 + Q^3 - 27\\ \pi^* &= 68.31 \text{ (\$ per day)}\end{aligned}$$

Example 5.5 In a perfect competition market structure, all firms are price-takers since they treat price as a market-determined constant. Digital Solutions Company's total revenue function is

$$\text{TR}(Q) = PQ$$

Assume that $P = 6$ and the total cost function is

$$\text{TC}(Q) = -2Q^3 + 45Q^2 - 330Q - 21$$

Determine the firm's profit function and the level of output at which Digital Solutions Company should produce in order to maximize profits.

Solution: The profit function of the firm is

$$\pi = \text{TR} - \text{TC} = 2Q^3 - 45Q^2 + 336Q + 21$$

where Q is in units (say, pounds per day).

The firm's optimal level of output can be found by using the first-order condition:

$$\frac{d\pi}{dQ} = 6Q^2 - 90Q + 336 = 0 \quad \text{when } Q = \begin{cases} 7 \\ 8 \end{cases}$$

The second derivative of the profit function is

$$\frac{d^2\pi}{dQ^2} = 12Q - 90 \quad \begin{cases} > 0 & \text{when } Q = 8 \\ < 0 & \text{when } Q = 7 \end{cases}$$

Thus, the profit-maximizing output level is $Q^* = 7$ (pounds per day). Note that at $Q = 8$ units level of production the profit is minimized. The maximized profit is found as

$$\pi^* = 2 \cdot 7^3 - 45 \cdot 7^2 + 336 \cdot 7 + 21 = 854 \text{ (\$ per day)}$$

Example 5.6 Consider a perfectly competitve firm with capital stock fixed at $K = \overline{K}$ in the short run. The production function of the firm is

$$Q = F(\overline{K}, L) = F(L)$$

where Q and L are the firm's output and labor. Suppose that

$$\frac{dQ}{dL} = F_L > 0$$

$$\frac{d^2Q}{dL^2} = F_{LL} < 0$$

Here the marginal product of labor is positive, but is diminishing. The firm is a price-taker and the price of the product it sells is P, the rental cost of capital is r and the wage rate is w. The firm chooses its labor input L to maximize profits. The firm's profits π are given by

$$\pi = PF(L) - rK - wL$$

The first-order condition for profit maximization is

$$\frac{d\pi}{dL} = PF_L - w = 0$$

which implies that

$$PF_L = w$$

Note that the second-order condition for profit maximization is satisfied:

$$\frac{d^2\pi}{dL^2} = PF_{LL} < 0$$

Thus, the firm increases its labor input to the point where the value of marginal product equals wage.

EXAMPLE 5.7 Consider a firm that uses only labor as its input and has the production function

$$f(L) = \alpha L^{\beta}$$

where L is input, $\alpha > 0$ and $0 < \beta < 1$. The firm's profit π is

$$\pi = Pf(L) - wL$$

where P is the price of the firm's product and w is the wage rate.

Find the profit-maximizing number of workers the firm will hire.

Solution: The firm's profit function is

$$\pi = \pi(L) = P\alpha L^{\beta} - wL$$

The first-order condition for profit maximization is

$$\frac{d\pi}{dL} = P\alpha\beta L^{\beta-1} - w = 0$$

It follows that the optimal amount of labor L^* is

$$L^* = \left(\frac{w}{P\alpha\beta}\right)^{\frac{1}{\beta-1}}$$

Note that the second-order condition for profit maximization holds:

$$\frac{d^2\pi}{dL^2} = P\alpha\beta(\beta - 1)L^{\beta-2} < 0$$

under the assumption that $\alpha,\ P > 0$ and $0 < \beta < 1$.

EXAMPLE 5.8 Consider a firm that uses labor as its only input and has the production function

$$f(L) = L^{\frac{1}{2}}$$

where L is the labor input. The firm's profit π is

$$\pi = Pf(L) - wL$$

where P is the price of the firm's product and w is the wage rate. Find the profit-maximizing number of workers the firm will hire.

Solution: The firm's profit function is

$$\pi = \pi(L) = PL^{\frac{1}{2}} - wL$$

The first-order condition for profit maximization is

$$\frac{d\pi}{dL} = \frac{P}{2}L^{-\frac{1}{2}} - w = 0$$

It follows that the optimal amount of labor L^* is

$$L^* = \left(\frac{w}{0.5P}\right)^{-2}$$

Note that the second-order condition for profit maximization holds:

$$\frac{d^2\pi}{dL^2} = \frac{P}{2}\left(\frac{1}{2} - 1\right)L^{-\frac{3}{2}} < 0$$

under the assumption that $P > 0$.

For a variety of reasons, economists differentiate between the following types of functions:

(a) functions increasing at an increasing rate;
(b) functions increasing at a decreasing rate;
(c) functions decreasing at an increasing rate;
(d) functions decreasing at a decreasing rate.

Those functions together with their characteristics are listed in Table 5.1.

The mathematical way of describing the type of behavior we observe in these examples is the notion of concave and convex functions. Functions of type (a) and (d), as depicted in Figure 5.5, are called *convex*. Functions of type (b) and (c) are called *concave*.

As Table 5.1 demonstrates, this behavior may be classified by the second derivative of a function. If the second derivative of a function is positive on an interval, then the function is *strictly convex*. If the second derivative is negative, then the function is *strictly concave*.

Table 5.1

	$f'(x)$	$f''(x)$	$f(x)$
(a)	>0	>0	increasing at an increasing rate
(b)	>0	<0	increasing at a decreasing rate
(c)	<0	<0	decreasing at an increasing rate
(d)	<0	>0	decreasing at a decreasing rate

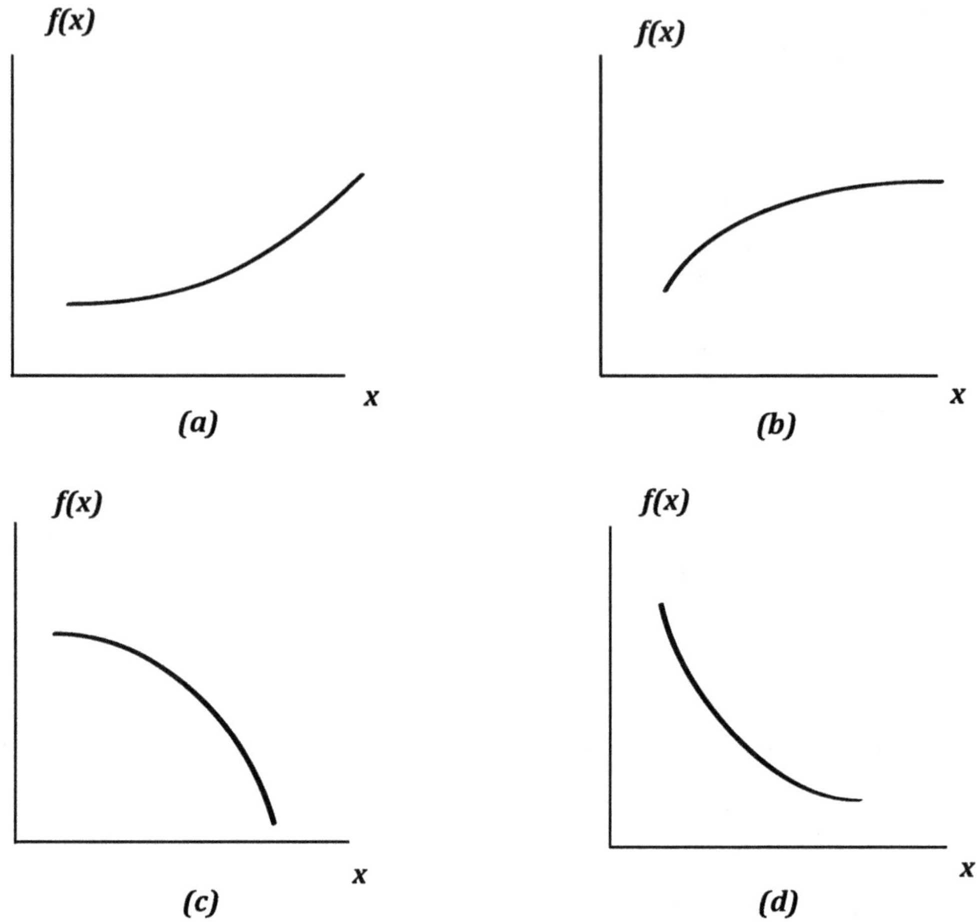

Figure 5.5 Illustration of Table 5.1.

If the second derivative is non-negative, we use the term *convex function*. If the second derivative is non-positive, then the function is *concave*.

Example 5.9 Find the intervals where $f(x) = 3x^4 - 4x^3 - 12x^2 + 5$ is increasing and decreasing.

Solution: The first derivative of $f(x)$ is

$$f'(x) = 12x^3 - 12x^2 - 24x = 12x(x-2)(x+1)$$

The sign of $f'(x)$ depends on the signs of the three factors of $f'(x)$, namely, $12x$, $(x-2)$ and $(x+1)$. We divide the real line into intervals whose endpoints are the critical points $-1, 0$ and 2, and arrange our work in a table, such as Table 5.2. A plus sign indicates that

Table 5.2

Interval	$12x$	$x-2$	$x+1$	$f'(x)$	f
$x<-1$	$-$	$-$	$-$	$-$	decr. on $(-\infty,-1)$
$-1<x<0$	$-$	$-$	$+$	$+$	incr. on $(-1,0)$
$0<x<2$	$+$	$-$	$+$	$-$	decr. on $(0,2)$
$x>2$	$+$	$+$	$+$	$+$	incr. on $(2,\infty)$

the expression in question is positive, and a minus sign indicates that it is negative. The last column of Table 5.2 provides us with the final conclusion.

EXAMPLE 5.10 Discuss the curve $y = x^4 - 4x^3$with respect to concavity, points of inflection, and local maxima and minima. Use this information to sketch the curve.

Solution: Let $y = f(x) = x^4 - 4x^3$, then

$$f'(x) = 4x^3 - 12x^2 = 4x^2(x-3)$$
$$f''(x) = 12x^2 - 24x = 12x(x-2)$$

To find the critical points we set $f'(x) = 0$ and obtain $x = 0$ and $x = 3$. Next, we evaluate $f''(x)$ at these critical points:

$$f''(0) = 0 \quad \text{and} \quad f''(3) = 36 > 0$$

Since $f'(3) = 0$ and $f''(3) > 0$, we conclude that point $A = (3, -27)$ is a local minimum. Since $f''(0) = 0$, the second-derivative test gives us no information about the nature of the critical point $x = 0$. But since $f'(x) < 0$ in the neigborhood of $x = 0$, the first-derivative test tells us that $f(x)$ does not have a local maximum or minimum at point 0. In fact, the expression for $f'(x)$ shows that f decreases to the left of $x = 3$ and to the right of $x = 3$.

Since $f''(x) = 0$ when $x = 0$ or 2, we divide the real line into intervals with these numbers as endpoints and present estimation results, as in Table 5.3.

Table 5.3

Interval	$f''(x) = 12x(x-2)$	*Curvature*
$(-\infty, 0)$	$+$	convex
$(0, 2)$	$-$	concave
$(2, \infty)$	$+$	convex

The point $(0, 0)$ is an inflection point since the function changes its curvature from convex to concave. Also $(2, 16)$ is an inflection point since the curvature of the function changes from concave to convex. Using our results on local extrema, inflection points and curvature, we sketch the graph in Figure 5.6.

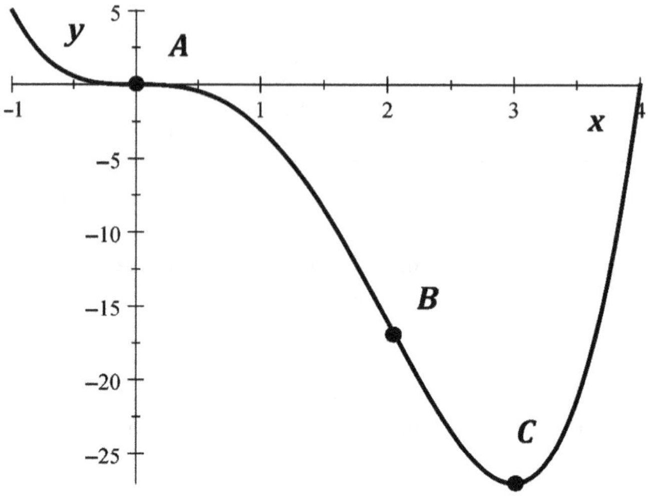

Figure 5.6

PROBLEMS FOR SECTION 5.1

1. Find the stationary points for the following and check whether they are maxima, minima or inflection points.

   (a) $y = (x-3)^2 + 5$
   (b) $y = (x-6)^2 + 7$
   (c) $y = (x+5)^2 - 4$
   (d) $y = 2x^2 - x + 17$
   (e) $y = -3x^2 - 6x + 12$

2. Find the stationary points for the following and check whether they are maxima, minima or inflection points.

   (a) $y = \dfrac{x^3}{3} - 5x^2 + 21x + 26$
   (b) $y = x^3 - 7.5x^2 + 18x + 25$
   (c) $y = x^3 + 8x^2 - 35x - 2$
   (d) $y = \dfrac{x^3}{3} - \dfrac{5x^2}{2} - 50x + 10$
   (e) $y = 11 + 121x - 22x^2 + x^3$

3. Consider a monopolist's linear demand function

$$P = 11 - 0.5Q$$

where P is the price of the good and $Q \in [0, 6]$ is its quantity. The monopolist's total cost function is

$$\text{TC} = -\frac{Q^3}{3} + 3Q^2 + Q + 23$$

Determine the level of output at which the monopolist should produce in order to maximize her profits. What is the optimal price that corresponds to this quantity?

4. Let the total revenue and total cost functions be

$$\text{TR}(Q) = 2.5Q - 2.25Q^2$$
$$\text{TC}(Q) = -80Q + 18Q^2 - Q^3 + 27$$

where TR and TC are in dollars and Q is in units (say, pounds per day). Find the profit-maximizing quantity of the output.

5. Consider a monopolist's constant elasticity demand curve:

$$P = bQ^{-\alpha}$$

where P is the price of the good and Q is its quantity. The monopolist's total cost function is

$$\text{TC} = aQ$$

Determine the first- and second-order conditions to maximize the monopolist's profits. When does the first-order condition hold?

6. The Kool Kola company is facing fierce competition from ColaStar. The management of Kool Kola is trying to increase the company's profit by minimizing total costs. Kool Kola total cost function is

$$\text{TC}(Q) = -\frac{1}{3}Q^3 + 10Q^2 - 100Q - 28$$

and the total revenue funciton is

$$\text{TR}(Q) = 5Q - Q^2$$

What is the cost-minimizing output level for Kool Kola? What is the profit-maximizing output level? As a manager, should you choose the profit-maximizing output level or cost-minimizing output level?

7. In a perfect competition market structure, all firms are price-takers since they treat price as a market-determined constant. Digital Solutions Company's total revenue function is

$$\text{TR}(Q) = PQ$$

in which P equals the output price. Assume that $P = 8$ and the total cost function is

$$\text{TC}(Q) = -2Q^3 + 21Q^2 - 64Q - 27$$

(a) Determine the firm's profit function and the level of output at which Digital Solutions Company should produce in order to maximize profits.
(b) According to microeconomic theory, perfectly competitive firms will maximize profits by producing at the quantity where price equals marginal cost and where the

slope of the marginal cost curve is positive. Demonstrate that the theory holds in this example.

8. Find the critical points of the following.

   (a) $y = (x-4)^5 + 16$
   (b) $y = (x-8)^3$
   (c) $y = (5-x)^4 + 3$

   Use the higher-order derivative test to determine whether they represent local extrema or inflection points.

5.2 Taylor series

Suppose we are working with a function $f(x)$ that is continuous and smooth enough on an interval about $x = a$. We can approximate $f(x)$ near point a by a polynomial $P_n(x)$ of degree n:

$$P_n(x) = f(a) + \frac{f'(a)}{1!}(x-a) + \cdots + \frac{f^{(n)}(a)}{n!}(x-a)^n$$

This is known as the Taylor polynomial of $f(x)$ of degree n at point a. When $a = 0$, it is usually called the Maclaurin polynomial of $f(x)$ of degree n.

THEOREM 5.4: (TAYLOR'S THEOREM). *Let f be a real function that has $n+1$ continuous derivatives on an open interval I containing point a. Then for any x in I, we have*

$$f(x) = \sum_{k=0}^{n} \frac{f^{(k)}(a)}{k!}(x-a)^k + R_n(x) = P_n(x) + R_n(x)$$

Let us comment on Theorem 5.4. The zeroth derivative of $f(x)$ is, by definition, $f(x)$ itself. Function $R_n(x)$ is the so-called remainder term. It is assumed to be insignificant when $n \to \infty$. There are several results that specify more precisely what $R_n(x)$ is, for instance,

$$R_n(x) = \frac{f^{(n+1)}(z)}{(n+1)!}(x-a)^{n+1}$$

where z is some number between x and a. For most functions, we cannot really say much more about the remainder term. However, for the "good" case of the so-called analytical functions, one can demonstrate that $R_n(x)$ goes to zero as n increases. The real meaning of Taylor's theorem for our applications is that the function approximately equals its Taylor polynomial:

$$f(x) \approx \sum_{k=0}^{n} \frac{f^{(k)}(a)}{k!}(x-a)^k$$

The above is the form of Taylor's theorem we use for the rest of the book. In the particular case when $a = 0$, the series is also called the Maclaurin series.

EXAMPLE 5.11 Find the Taylor polynomial of degree five for the function

$$f(x) = e^x$$

around point $a = 1$.

Solution: Note that

$$\begin{aligned} f^{(1)}(x) &= e^x \\ f^{(2)}(x) &= e^x \\ f^{(3)}(x) &= e^x \\ f^{(4)}(x) &= e^x \\ f^{(5)}(x) &= e^x \end{aligned}$$

and $f(1) = e$. Thus,

$$f(x) \approx e + e(x-1) + \frac{e(x-1)^2}{2!} + \frac{e(x-1)^3}{3!} + \frac{e(x-1)^4}{4!} + \frac{e(x-1)^5}{5!}$$

EXAMPLE 5.12 Compute the first- and second-degree Taylor polynomials for the exponential function $f(x) = e^x$ around $x = 1$.

Solution: Since

$$f^{(1)}(1) = f^{(2)}(1) = e$$

we have that

$$\begin{aligned} P_1(x) &= e + e(x-1) = ex \\ P_2(x) &= ex + \frac{e}{2}(x-1)^2 \end{aligned}$$

Note that

$$P_1(1) = e \quad \text{and} \quad P_2(1) = e$$

which illustrates that the first- and second-order Taylor polynomials evaluated at $x = a$ produce the exact and not approximate value of the function $f(x)$ at $x = a$. The reader should have no problem showing that

$$P_n(a) = f(a) \quad \forall\, n$$

In general, a function $f(x)$ can be approximated about $x = a$ with nth-degree Taylor polynomial as follows:

$$f(x) \approx f(a) + \frac{f'(a)}{1!}(x-a) + \cdots + \frac{f^{(n)}(a)}{n!}(x-a)^n$$

The approximation of $f(x)$ by the Taylor polynomial of degree one is also known as the linear approximation of $f(x)$ about $x = a$. Thus, the linear approximation of $f(x)$ about $x = a$ is

$$f(x) \approx f(a) + f'(a)(x - a) \tag{5.1}$$

Often economists replace complicated functions with linear ones in order to simplify matters.

EXAMPLE 5.13 Find the linear approximation to $f(x) = x^2$ about $a = 1$.

Solution: Using equation (5.1) yields

$$f(x) = 1 + 2(x - 1)$$

So, we have

$$f(x) \approx 2x - 1$$

EXAMPLE 5.14 Use linear approximation to find an approximate value for $(1.002)^{100}$.

Solution: Let us use the function $f(x) = x^{100}$. Then $f(1) = 1$ and $f'(1) = 100 \cdot 1^{99} = 100$. Thus, by formula (5.1)

$$\begin{aligned}(1.02)^{100} &\approx 1 + 100(1.002 - 1)\\ &= 1.2\end{aligned}$$

Using Mathematica we find (1.002)^100 ≈ 1.2212.

EXAMPLE 5.15 Find the linear approximation to the following functions about $x = 1$.

(a) $f(x) = (x + 2)^3$
(b) $f(x) = (x + 3)^{\frac{1}{2}}$
(c) $f(x) = \dfrac{1}{x + 1}$
(d) $f(x) = \ln(x)$

Solution:

(a) Here $f'(x) = 3(x + 2)^2$ and $f'(1) = 27$. Since $f(1) = 27$, using (5.1) we get

$$\begin{aligned}(x + 2)^3 &\approx 27 + 27(x - 1)\\ &= 27x\end{aligned}$$

(b) Here $f'(x) = \frac{1}{2}(x+3)^{-\frac{1}{2}}$ and $f'(1) = \frac{1}{4}$. Notice that $f(1) = 2$. From the linear approximation formula (5.1) we get

$$
\begin{aligned}
(x+3)^{\frac{1}{2}} &\approx 2 + \frac{1}{4}(x-1)\\
&= \frac{7+x}{4}
\end{aligned}
$$

(c) Here $f'(x) = -1(x+1)^{-2}$ and$f'(1) = -\frac{1}{4}$. The value of function $f(x)$ at $x = 1$ is $\frac{1}{2}$. Thus using (1) we find

$$
\begin{aligned}
\frac{1}{x+1} &\approx \frac{1}{2} - \frac{1}{4}(x-1)\\
&= \frac{3-x}{4}
\end{aligned}
$$

(d) Here $f'(x) = \dfrac{1}{x}$ and $f'(1) = 1$. Since $f(1) = 0$, using (5.1) we arrive at

$$\ln(x) \approx (x-1)$$

Approximation of order two is called the quadratic approximation, which can be found as follows:

$$f(x) \approx f(a) + f'(a)(x-a) + \frac{f''(a)}{2}(x-a)^2 \tag{5.2}$$

EXAMPLE 5.16 Find the quadratic approximation to the following functions about $x = 0$

(a) $f(x) = (x+2)^3$
(b) $f(x) = (x+3)^{\frac{1}{2}}$
(c) $f(x) = \dfrac{1}{x+1}$
(d) $f(x) = (2x-3)^2$

Solution:

(a) Here $f'(x) = 3(x+2)^2$, $f''(x) = 6(x+2)$ and so $f'(0) = 12$, $f''(0) = 12$. Because $f(0) = 8$, using the quadratic approximation formula we get

$$
\begin{aligned}
(x+2)^3 &\approx f(0) + f'(0)(x-0) + \frac{1}{2}f''(0)(x-0)^2\\
&= 8 + 12x + 6x^2
\end{aligned}
$$

(b) Here $f'(x) = \frac{1}{2}(x+3)^{-\frac{1}{2}}$, $f''(x) = -\frac{1}{4}(x+3)^{-\frac{3}{2}}$ and so $f'(0) = \frac{\sqrt{3}}{6}$, $f''(0) = \frac{-\sqrt{3}}{36}$. Because $f(0) = \sqrt{3}$, using formula (2) we arrive at

$$(x+3)^{\frac{1}{2}} \approx f(0) + f'(0)(x-0) + \frac{1}{2}f''(0)(x-0)^2$$
$$= \sqrt{3} + \frac{\sqrt{3}}{6}x - \frac{\sqrt{3}}{72}x^2$$

(c) Here $f'(x) = -1(x+1)^{-2}$, $f''(x) = 2(x+1)^{-3}$ and so $f'(0) = -1$, $f''(0) = 2$. Since $f(0) = 1$, from the formula we obtain

$$\frac{1}{x+1} \approx f(0) + f'(0)(x-0) + \frac{1}{2}f''(0)(x-0)^2$$
$$= 1 - x + x^2$$

(d) Here $f'(x) = 8x - 12$, $f''(x) = 8$ and so $f'(0) = -12$, $f''(0) = 8$. It can be easily found that $f(0) = 9$. Then, the quadratic approximation formula yields

$$(2x-3)^2 \approx f(0) + f'(0)(x-0) + \frac{1}{2}f''(0)(x-0)^2$$
$$= 9 - 12x + 4x^2$$

As this example shows, for a second-degree polynomial its second-order Taylor approximation coincides with the polynomial.

EXAMPLE 5.17 Find the third-order Taylor approximation of the following function about $x = 0$.

$$f(x) = (x+1)^8$$

Solution: For the function $f(x) = (x+1)^8$ we have

$$f'(x) = 8(x+1)^7$$
$$f''(x) = 56(x+1)^6$$
$$f'''(x) = 336(x+1)^5$$

and so $f'(0) = 8$, $f''(0) = 56$ and $f'''(0) = 336$. Since $f(0) = 1$, using the Taylor approximation formula we obtain

$$(x+1)^8 \approx f(0) + f'(0)(x-0) + \frac{1}{2}f''(0)(x-0)^2 + \frac{1}{6}f'''(0)(x-0)^3$$
$$= 1 + 8x + 28x^2 + 56x^3$$

EXAMPLE 5.18 Find the fourth-order Taylor approximation of the following function about $x = 0$.

$$f(x) = (x+1)^8$$

Solution: For $f(x) = (x+1)^8$ we have

$$\begin{aligned} f'(x) &= 8(x+1)^7 \\ f''(x) &= 56(x+1)^6 \\ f^{(3)}(x) &= 336(x+1)^5 \\ f^{(4)}(x) &= 1680(x+1)^4 \end{aligned}$$

and so $f'(0) = 8$, $f''(0) = 56$, $f^{(3)}(0) = 336$ and $f^{(4)}(0) = 1680$. Because $f(0) = 1$, using the Taylor approximation formula we get

$$\begin{aligned} (x+1)^8 &\approx f(0) + f'(0)(x-0) + \frac{1}{2}f''(0)(x-0)^2 + \frac{1}{6}f'''(0)(x-0)^3 + \frac{1}{24}f'''(0)(x-0)^4 \\ &= 1 + 8x + 28x^2 + 56x^3 + 70x^4 \end{aligned}$$

EXAMPLE 5.19 Find the fifth-order approximation to $f(x) = e^x$ about $x = 0$.

Solution: Since all derivatives of $f(x)$ are equal to e^x, we have

$$\begin{aligned} f'(0) &= 1 \\ f''(0) &= 1 \\ f^{(3)}(0) &= 1 \\ f^{(4)}(0) &= 1 \\ f^{(5)}(0) &= 1 \end{aligned}$$

or, alternatively, $f^{(n)}(0) = 1$ for all $n = 1, 2, 3, 4, 5$. Thus,

$$\begin{aligned} e^x &\approx 1 && (n=1) \\ e^x &\approx 1 + \frac{x}{1!} && (n=2) \\ e^x &\approx 1 + \frac{x}{1!} + \frac{x^2}{2!} && (n=3) \\ e^x &\approx 1 + \frac{x}{1!} + \frac{x^2}{2!} + \frac{x^3}{3!} && (n=4) \\ e^x &\approx 1 + \frac{x}{1!} + \frac{x^2}{2!} + \frac{x^3}{3!} + \frac{x^4}{4!} && (n=5) \\ e^x &\approx 1 + \frac{x}{1!} + \frac{x^2}{2!} + \frac{x^3}{3!} + \frac{x^4}{4!} + \frac{x^5}{5!} && (n=6) \end{aligned}$$

PROBLEMS FOR SECTION 5.2

1. Find the linear approximation of the function $f(x) = \sqrt{1-x}$ at $a = 0$ and use it to estimate the numbers $\sqrt{0.95}$ and $\sqrt{0.98}$.
2. Prove that

$$\ln(1+x) \approx \frac{x}{1+x}$$

for x close to 0 and draw the graphs of $y = \frac{x}{1+x}$ and $y = \ln(1+x)$ in the same coordinate system using Mathematica or Matlab. (Hint: use linear approximations.)

3. Find the linear approximation of the function $f(x) = \sqrt{x+8}$ at $a = 1$ and use it to estimate the numbers $\sqrt{8.99}$ and $\sqrt{9.02}$.
4. Find the linear approximations to the following functions about $x = 1$.

   (a) $f(x) = (x+5)^2$

   (b) $f(x) = (x+1)^{\frac{1}{2}}$

   (c) $f(x) = \frac{1}{2x+1}$

   (d) $f(x) = (2x-1)^3$

5. Find the linear approximation of the function $g(x) = \sqrt[3]{1+x}$ at $a = 0$ and use it to estimate the numbers $\sqrt[3]{0.97}$ and $\sqrt[3]{1.01}$.
6. Find the quadratic approximations to the following functions about $x = 0$.

   (a) $f(x) = (3+x)^4$

   (b) $f(x) = (x+4)^{\frac{1}{2}}$

   (c) $f(x) = \frac{1}{3x+1}$

   (d) $f(x) = (5x-1)^3$

7. Find the nth-degree Taylor approximation for $f(x) = e^{-x}$ about $a = 0$.
8. Find the Taylor polynomial of degree 3 about $a = 0$ for the following functions.

   (a) $f(x) = e^x + 1$

   (b) $(x) = \sqrt{9+x}$

   (c) $h(x) = (x+1)^5$

   (d) $w(x) = \ln(x+2)$

5.3 Mathematica examples

In the following example we model a production function as a function of labor only:

$$f(L) = L^{\frac{1}{2}}$$

The production function is given as follows:

```
>q = L^0.5
```

```
L^0.5
```

The objective function is the profit function given by the difference between total revenue and total cost:

```
>profit[L_] = P * q - w * L
```

```
L^0.5 P-Lw
```

To solve for the optimal level of labor usage, we find the first and second derivatives of the profit function:

> **fdprofit = profit′[L]**

$$\frac{0.5\text{P}}{\text{L}^{0.5}}-\text{w}$$

> **sdprofit = profit″[L]**

$$-\frac{0.25\text{P}}{\text{L}^{1.5}}$$

The stationary point is found by solving the first-order condition for L:

> **sol = Solve[{fdprofit == 0}, {L}]**

$$\left\{\left\{L\text{->}\frac{0.25}{\left(\frac{w}{P}\right)^{2.0}}\right\}\right\}$$

Note that L^* is positive. We verify that the second-order condition for profit maximization holds by observing that the second derivative of the profit function is always negative:

$$-\frac{0.25P}{L^{1.5}}<0$$

We can graph the production function $f(L)$ using the **Plot** command, as shown in Figure 5.7:

> **Plot[L^0.5, {L, 0, 100}]**

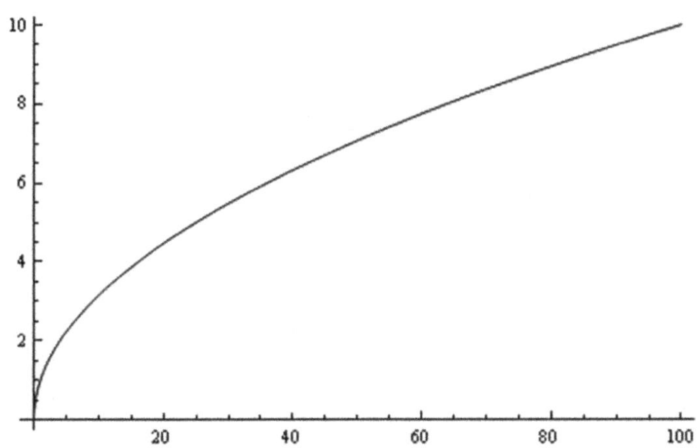

Figure 5.7

Mathematica allows us to graph the production function using a dashed line, as shown in Figure 5.8:

>Plot[L^0.5, {L, 0, 100}, PlotStyle− > Directive[Black, Dashed]]

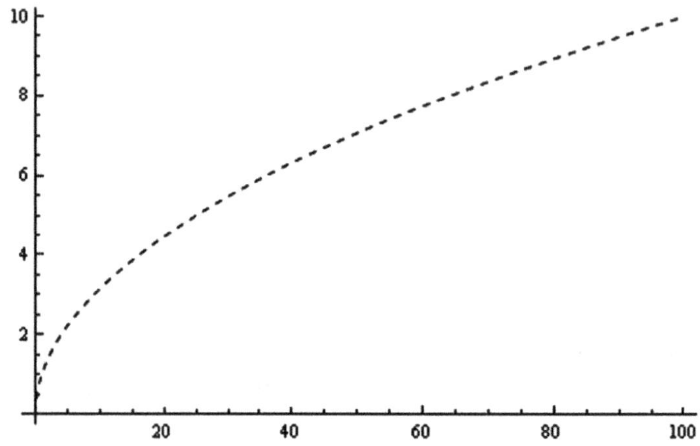

Figure 5.8

We can also graph the production function using a thick line, as shown in Figure 5.9:

>Plot[L^0.5, {L, 0, 100}, PlotStyle− > Directive[Black, Thick]]

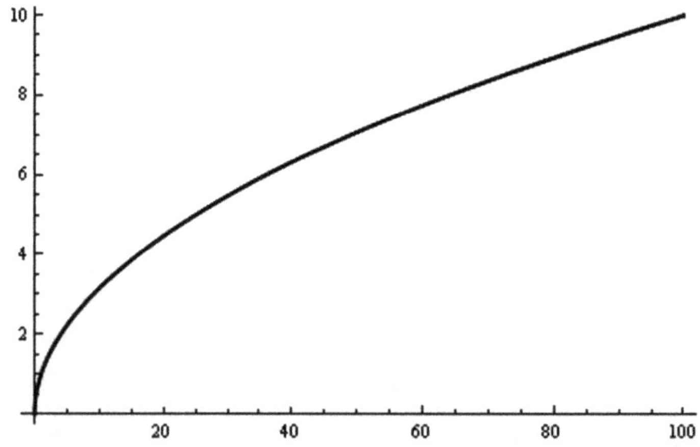

Figure 5.9

Let us set $P = 20$ and $w = 1$. Now, we shall plot the profit function (shown in Figure 5.10) and its first two derivatives (shown in Figure 5.11 and 5.12):

```
>{P = 20, w = 1};
>Plot[profit[L], {L, 0, 500}, PlotStyle- > Directive[Black, Thick]]
>Plot[profit'[L], {L, 0, 500}, PlotStyle- > Directive[Black, Thick]]
>Plot[profit''[L], {L, 0, 500}, PlotStyle- > Directive[Black, Thick]]
```

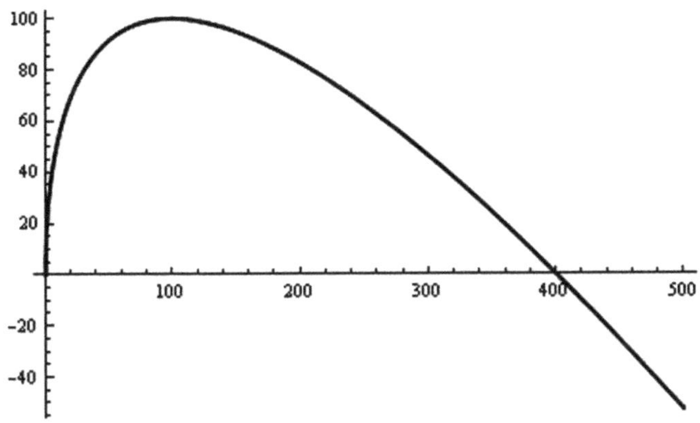

Figure 5.10 Profit function.

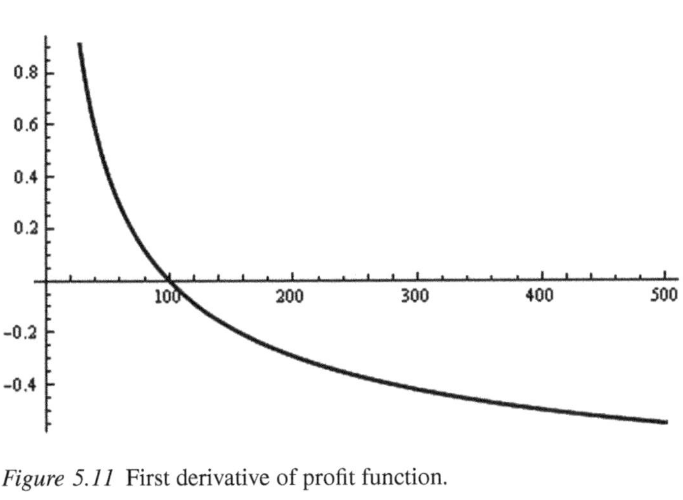

Figure 5.11 First derivative of profit function.

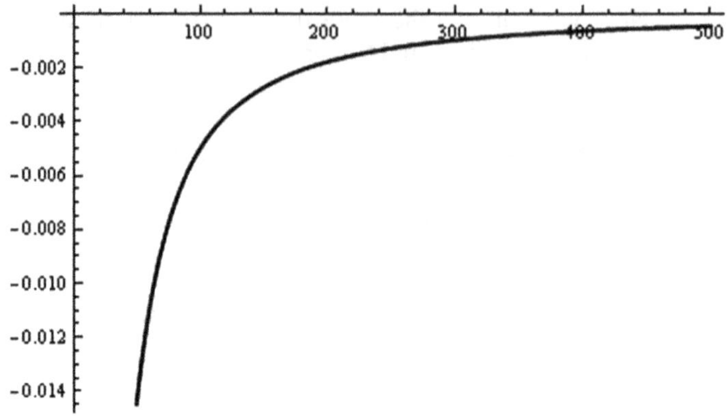

Figure 5.12 Second derivative of profit function.

Now we can clearly see that $a = 100$ is the critical point of the profit function, and the profit function attains a maximum at that point.

6 Matrix algebra

Matrices are essentially tables with numbers. As such, they are as frequent as the tables you create in applications like Microsoft Excel. Many business data are naturally organized as tables. For instance, columns may label various types of merchandise a firm has, while rows list various business metrics of the merchandise (metrics like cost, profit, labor per unit, etc.). However, unlike tables, which just store data, there are meaningful algebraic operations that we can apply to matrices. For instance, matrix multiplication naturally models such fundamental operations as totaling across certain parameters.

6.1 Introduction

A *matrix* is a rectangular array of elements. These elements may be numbers, parameters, variables or functions. A matrix that has m rows and n columns is said to have *dimension* $m \times n$ (read "m by n"). A matrix that has m rows and n columns sometimes is said to have *order* $m \times n$. A matrix of dimension $m \times n$ can be written in three forms:

$$\mathbf{A} = \begin{bmatrix} a_{11} & a_{12} & \dots & a_{1n} \\ a_{21} & a_{22} & \dots & a_{2n} \\ \dots & \dots & \dots & \dots \\ a_{m1} & a_{m2} & \dots & a_{mn} \end{bmatrix} = [a_{ij}]_{(m,n)} = [a_{ij}]$$

We can also write $\text{Dim}(\mathbf{A}) = m \times n$ to indicate that matrix $\mathbf{A}$ has a dimension of $m \times n$.

EXAMPLE 6.1 The array

$$\mathbf{A}_1 = \begin{bmatrix} 1 & 23 & -17 & 31 \\ 32 & 21 & 53 & 3 \\ 7 & 14 & 11 & 9 \end{bmatrix}$$

is called a 3×4 matrix.

A matrix that has m rows and n columns is said to have a *row dimension* of m and a *column dimension* of n. The location of each element a_{ij} in a matrix is given by two subscripts, i and j. For instance, the element a_{34} is located in the third row and fourth column.

Matrix $\mathbf{A}$ is said to be a *square matrix* if $\text{Dim}(\mathbf{A}) = n \times n$. In other words, a matrix is square if the number of its rows is equal to the number of its columns.

EXAMPLE 6.2 Matrices

$$\mathbf{A}_2 = \begin{bmatrix} 32 & 1 & 44 & 543 \\ 76 & 11 & 43 & 32 \\ 56 & 236 & 65 & 76 \\ 543 & 34 & 521 & 64 \end{bmatrix}$$

and

$$\mathbf{A}_3 = \begin{bmatrix} 1 & 32 & 51 \\ 6 & 8 & 32 \\ 74 & 5 & 12 \end{bmatrix}$$

are square matrices of dimension 4×4 and 3×3 respectively. Sometimes a square matrix of order $n \times n$ is referred to as a square matrix of order n.

The *main diagonal* of a matrix consists of elements for which i equals j. For instance, the main diagonal of matrix $\mathbf{A}_1$ in Example 6.1 consists of elements 1, 21 and 11. The main diagonal of a square matrix is the diagonal that runs from the top left corner to the bottom right corner. Note that the main diagonal of matrix $\mathbf{A}_3$ in Example 6.2 consists of elements 1, 8 and 12.

Matrices **A** and **B** are *equal if* and only if

$$\text{Dim}(\mathbf{A}) = \text{Dim}(\mathbf{B})$$

and

$$a_{ij} = b_{ij} \quad \forall\, i \text{ and } j$$

EXAMPLE 6.3 Let **A** and **B** be two matrices of dimension 3×4, where

$$\mathbf{A} = \begin{bmatrix} 2 & 1 & 5 & 7 \\ 11 & 51 & 3 & 4 \\ 234 & 32 & 1 & 14 \end{bmatrix}$$

and

$$\mathbf{B} = \begin{bmatrix} 2 & 1 & 5 & 7 \\ 11 & 51 & 3 & 4 \\ 234 & 32 & 1 & 14 \end{bmatrix}$$

Then we can write $\mathbf{A} = \mathbf{B}$ to indicate that matrix **A** is equal to matrix **B**.

A matrix can be multiplied by a scalar and the operation itself is called *scalar multiplication*. Given a scalar k and a matrix **A**, then

$$k\mathbf{A} = [ka_{ij}]$$

EXAMPLE 6.4

$$5\begin{bmatrix} 2 & 4 & 12 & 3 \\ 8 & 3 & 6 & 1 \\ 11 & 4 & -5 & 15 \end{bmatrix} = \begin{bmatrix} 10 & 20 & 60 & 15 \\ 40 & 15 & 30 & 5 \\ 55 & 20 & -25 & 75 \end{bmatrix}$$

and

$$-2\begin{bmatrix} 2 & 4 & 12 & 3 \\ 8 & 3 & 6 & 1 \\ 11 & 4 & -5 & 15 \end{bmatrix} = \begin{bmatrix} -4 & -8 & -24 & -6 \\ -16 & -6 & -12 & -2 \\ -22 & -8 & 10 & -30 \end{bmatrix}$$

6.1.1 Special matrices

A square matrix of dimension $n \times n$ with ones on the main diagonal and zeros everywhere else is called the identity matrix and denoted $\mathbf{I}_n$.

EXAMPLE 6.5

$$\mathbf{I}_1 = [1]$$

$$\mathbf{I}_2 = \begin{bmatrix} 1 & 0 \\ 0 & 1 \end{bmatrix}$$

$$\mathbf{I}_n = \begin{bmatrix} 1 & 0 & \dots & 0 \\ 0 & 1 & \dots & 0 \\ & & \dots & \\ 0 & 0 & \dots & 1 \end{bmatrix}$$

Sometimes it is convenient to refer to a matrix of dimension $m \times n$ as $\mathbf{A}_{m\times n}$. If all elements of matrix $\mathbf{A}_{m\times n}$ are zero, the matrix is called the *null matrix* and is denoted by $\mathbf{0}_{m\times n}$.

EXAMPLE 6.6

$$\mathbf{0}_{3\times 4} = \begin{bmatrix} 0 & 0 & 0 & 0 \\ 0 & 0 & 0 & 0 \\ 0 & 0 & 0 & 0 \end{bmatrix}$$

A square matrix is *upper triangular* if all elements below the main diagonal are zero. A square matrix is *lower triangular* if all elements above the main diagonal are zero. A square matrix is *triangular* if it is upper triangular or lower triangular. The following matrix is upper triangular:

$$\mathbf{B} = \begin{bmatrix} 21 & 5 & 8 & 98 \\ 0 & 12 & 3 & 54 \\ 0 & 0 & 322 & 33 \\ 0 & 0 & 0 & 11 \end{bmatrix}$$

A *diagonal matrix* is a square matrix in which the entries outside the main diagonal are all zero. The diagonal entries themselves may or may not be zero.

EXAMPLE 6.7 The matrix

$$\mathbf{A} = \begin{bmatrix} 2 & 0 & 0 & 0 \\ 0 & 23 & 0 & 0 \\ 0 & 0 & 7 & 0 \\ 0 & 0 & 0 & 4 \end{bmatrix}$$

is a diagonal matrix.

A *row vector* is a matrix of dimension $1 \times n$. A *column vector* is a matrix of dimension $n \times 1$. We will use lower-case letters to denote row or column vectors.

EXAMPLE 6.8

$$\mathbf{a} = \begin{bmatrix} 21 & 43 & 11 & 15 \end{bmatrix}$$

is a row vector and

$$\mathbf{b} = \begin{bmatrix} 14 \\ 5 \\ 3 \\ 67 \\ 23 \end{bmatrix}$$

is a column vector.

6.1.2 Matrix addition and subtraction

Two matrices of the same dimension are said to be *conformable for addition or subtraction*. Only matrices of the same dimension can be added or subtracted. Given two conformable matrices **A** and **B**,

$$\mathbf{C} = \mathbf{A} + \mathbf{B} = [a_{ij} + b_{ij}]$$
$$\mathbf{D} = \mathbf{A} - \mathbf{B} = [a_{ij} - b_{ij}]$$

EXAMPLE 6.9

$$\begin{bmatrix} 2 & 1 & 5 \\ 4 & 7 & 11 \end{bmatrix} + \begin{bmatrix} 3 & 1 & 5 \\ 81 & 2 & 4 \end{bmatrix} = \begin{bmatrix} (2+3) & (1+1) & (5+5) \\ (4+81) & (7+2) & (11+4) \end{bmatrix}$$
$$= \begin{bmatrix} 5 & 2 & 10 \\ 85 & 9 & 15 \end{bmatrix}$$

and

$$\begin{bmatrix} 2 & 1 & 5 \\ 4 & 7 & 11 \end{bmatrix} - \begin{bmatrix} 3 & 1 & 5 \\ 81 & 2 & 4 \end{bmatrix} = \begin{bmatrix} (2-3) & (1-1) & (5-5) \\ (4-81) & (7-2) & (11-4) \end{bmatrix}$$

$$= \begin{bmatrix} -1 & 0 & 0 \\ -77 & 5 & 7 \end{bmatrix}$$

Matrix addition is commutative

$$\mathbf{A} + \mathbf{B} = \mathbf{B} + \mathbf{A}$$

and associative

$$(\mathbf{A} + \mathbf{B}) + \mathbf{C} = \mathbf{A} + (\mathbf{B} + \mathbf{C})$$

In addition,

$$\mathbf{A} + \mathbf{0} = \mathbf{A}$$

and

$$\mathbf{A} + (-\mathbf{A}) = \mathbf{0}$$

6.1.3 Matrix multiplication

Matrix multiplication is one of the most important operations in matrix algebra and can only be performed on matrices that are conformable for multiplication. Matrices **A** and **B** are *conformable for multiplication* if the number of columns of matrix **A** is equal to the number of rows of matrix **B**. Generally, matrices $\mathbf{A}_{m \times n}$ and $\mathbf{B}_{n \times p}$ are conformable for multiplication and the resulting matrix **C** has a dimension of $m \times p$.

EXAMPLE 6.10 Suppose

$$\mathbf{A} = \begin{bmatrix} 21 & 5 & 8 & 98 \\ 0 & 12 & 3 & 54 \\ 0 & 0 & 322 & 33 \\ 0 & 0 & 0 & 11 \end{bmatrix}$$

and

$$\mathbf{B} = \begin{bmatrix} 1 & 2 \\ 5 & 7 \\ 15 & 21 \\ 24 & 63 \end{bmatrix}$$

then **A** and **B** are conformable for multiplication. In addition,

$$\text{Dim}(\mathbf{AB}) = 4 \times 2$$

Sometimes it is convenient to refer to a specific row or column of a matrix $\mathbf{A}$. Let $\mathbf{a}^i$ denote the ith row of matrix $\mathbf{A}_{m\times n}$ and let $\mathbf{a}_i$ denote the ith column of matrix $\mathbf{A}_{m\times n}$. Similarly, let $\mathbf{b}^i$ denote the ith row of matrix $\mathbf{B}_{n\times p}$ and let $\mathbf{b}_i$ denote the ith column of matrix $\mathbf{B}_{n\times p}$.

EXAMPLE 6.11 Consider matrices **A** and **B** from Example 6.10. Then

$$\mathbf{a}^3 = \begin{bmatrix} 0 & 0 & 322 & 33 \end{bmatrix}$$

and

$$\mathbf{b}_2 = \begin{bmatrix} 2 \\ 7 \\ 21 \\ 63 \end{bmatrix}$$

Let $\mathbf{C}$ be the product of two conformable matrices: $\mathbf{A}_{m\times n}$ and $\mathbf{B}_{n\times p}$. Then the dimension of matrix $\mathbf{C}$ is $m \times p$ and

$$c_{ij} = \mathbf{a}^i \mathbf{b}_j$$

where $\mathbf{a}^i$ is the ith row of matrix $\mathbf{A}_{m\times n}$ and $\mathbf{b}_j$ is the jth column of matrix $\mathbf{B}_{n\times p}$. Thus, we can write

$$\mathbf{C} = \left[\mathbf{a}^i \mathbf{b}_j\right]_{m\times p} = \left[\sum_{k=1}^{n} a_{ik} b_{kj}\right]_{m\times p}$$

EXAMPLE 6.12 Suppose

$$\mathbf{A} = \begin{bmatrix} 3 & 7 & 1 & -2 \\ 5 & 1 & -3 & 4 \end{bmatrix}$$

and

$$\mathbf{B} = \begin{bmatrix} 3 & 5 \\ 2 & 4 \\ 8 & 1 \\ 7 & 2 \end{bmatrix}$$

then

$$\begin{aligned}
\mathbf{AB} &= \begin{bmatrix} 3 & 7 & 1 & -2 \\ 5 & 1 & -3 & 4 \end{bmatrix} \begin{bmatrix} 3 & 5 \\ 2 & 4 \\ 8 & 1 \\ 7 & 2 \end{bmatrix} \\
&= \begin{bmatrix} 3(3)+7(2)+1(8)+(-2)(7) & 3(5)+7(4)+1(1)+(-2)(2) \\ 5(3)+1(2)+(-3)(8)+4(7) & 5(5)+1(4)+(-3)(1)+4(2) \end{bmatrix} \\
&= \begin{bmatrix} 17 & 40 \\ 21 & 34 \end{bmatrix}
\end{aligned}$$

and

$$\mathbf{BA} = \begin{bmatrix} 3 & 5 \\ 2 & 4 \\ 8 & 1 \\ 7 & 2 \end{bmatrix} \begin{bmatrix} 3 & 7 & 1 & -2 \\ 5 & 1 & -3 & 4 \end{bmatrix}$$

$$= \begin{bmatrix} 34 & 26 & -12 & 14 \\ 26 & 18 & -10 & 12 \\ 29 & 57 & 5 & -12 \\ 31 & 51 & 1 & -6 \end{bmatrix}$$

Observe that **AB** is a 2×2 matrix and **BA** is a 4×4 matrix, which means that $\mathbf{AB} \neq \mathbf{BA}$. Generally, matrix multiplication is not commutative. In other words, in general **AB** is not equal to **BA**. In fact, the product **BA** may even be undefined.

EXAMPLE 6.13 Let

$$\mathbf{A} = \begin{bmatrix} 2 & 1 \\ 5 & 3 \\ 2 & 7 \\ 78 & 12 \end{bmatrix} \quad \text{and} \quad \mathbf{B} = \begin{bmatrix} 3 & 1 & 6 \\ 2 & 5 & 3 \end{bmatrix}$$

Is the product **AB** defined? Is the product **BA** defined?

Solution: **A** is a 4×2 matrix and **B** is a 2×3 matrix. The product **AB** is defined and has a dimension of 4×3. The product **BA** is undefined since the column dimension of **B** is not equal to the row dimension of **A**.

The following rules for conformable matrices hold:

(Mat1)	$(\mathbf{AB})\mathbf{C} = \mathbf{A}(\mathbf{BC})$	associative law
(Mat2)	$\mathbf{A}(\mathbf{B}+\mathbf{C}) = \mathbf{AB}+\mathbf{AC}$	left distributive law
(Mat3)	$(\mathbf{A}+\mathbf{B})\mathbf{C} = \mathbf{AC}+\mathbf{BC}$	right distributive law
(Mat4)	$\mathbf{A0} = \mathbf{0A} = \mathbf{0}$	mult. by null matrix
(Mat5)	$\mathbf{AI} = \mathbf{IA} = \mathbf{A}$	mult. by identity matrix

Looking at **Mat4** one may be tempted to think that $\mathbf{AB} = \mathbf{0}$ implies that either **A** or **B** is zero. However, this is not true in general.

EXAMPLE 6.14

$$\begin{bmatrix} -2 & 1 \\ -2 & 1 \end{bmatrix} \begin{bmatrix} 1 & 1 \\ 2 & 2 \end{bmatrix} = \begin{bmatrix} 0 & 0 \\ 0 & 0 \end{bmatrix}$$

Observe that the product of the two non-zero matrices is equal to the zero matrix.

6.1.4 Powers of a matrix

If $\mathbf{A}$ is a square matrix then we can write $\mathbf{A}$ multiplied by $\mathbf{A}$ as $\mathbf{A}^2$. Generally, $\mathbf{A}$ multiplied by itself k times is written as

$$\mathbf{A}^k = \mathbf{AA} \ldots \mathbf{A} \quad (k \text{ times})$$

EXAMPLE 6.15 Let

$$\mathbf{A} = \mathbf{I}_2$$

then

$$\mathbf{A}^2 = \mathbf{AA} = \begin{bmatrix} 1 & 0 \\ 0 & 1 \end{bmatrix} \begin{bmatrix} 1 & 0 \\ 0 & 1 \end{bmatrix}$$
$$= \begin{bmatrix} 1 & 0 \\ 0 & 1 \end{bmatrix} = \mathbf{A}$$

An identity matrix remains unchanged when multiplied by itself ($\mathbf{I}^2 = \mathbf{I}$). Any square matrix that remains unchanged when multiplied by itself ($\mathbf{A}^2 = \mathbf{A}$) is said to be *idempotent*.

6.1.5 Transpose of a matrix

The interchange of rows and columns of a matrix produces the transpose of that matrix. The transpose of matrix $\mathbf{A}$ is written as $\mathbf{A}'$. If the dimension of $\mathbf{A}$ is $m \times n$ then the dimension of $\mathbf{A}'$ is $n \times m$. Note that the row dimension of $\mathbf{A}$ is equal to the column dimension of $\mathbf{A}'$. The ith row of matrix $\mathbf{A}$ becomes the ith column of $\mathbf{A}'$ and

$$a_{ij} = a'_{ji}, \quad \text{where } i, j = 1, 2, \ldots, n$$

and a'_{ji} is the (j, i)th element of $\mathbf{A}'$.

EXAMPLE 6.16 Let

$$\mathbf{A} = \begin{bmatrix} 1 & 7 & 4 & 3 & 2 & 1 \\ 2 & -11 & 12 & 6 & 14 & 2 \\ 7 & 3 & 1 & 4 & 5 & -4 \end{bmatrix}$$

then

$$\mathbf{A}' = \begin{bmatrix} 1 & 2 & 7 \\ 7 & -11 & 3 \\ 4 & 12 & 1 \\ 3 & 6 & 4 \\ 2 & 14 & 5 \\ 1 & 2 & -4 \end{bmatrix}$$

EXAMPLE 6.17 Let

$$\mathbf{A} = \begin{bmatrix} a_{11} & a_{12} & a_{13} \\ a_{21} & a_{22} & a_{23} \end{bmatrix}$$

then

$$\mathbf{A}' = \begin{bmatrix} a_{11} & a_{21} \\ a_{12} & a_{22} \\ a_{13} & a_{23} \end{bmatrix}$$

If matrix **B** is the transpose of **A**, then the element a_{ij} of **A** becomes the element b_{ji} of **B**.

There are several important rules of transposition. If k is a scalar and **A** and **B** are matrices with

$$\text{Dim}(\mathbf{A}) = \text{Dim}(\mathbf{B}) = m \times n$$

then

$$\textbf{(Mat6)}\ (\mathbf{A}')' = \mathbf{A}$$

$$\textbf{(Mat7)}\ (\mathbf{kA})' = k\mathbf{A}'$$

$$\textbf{(Mat8)}\ (\mathbf{A} + \mathbf{B})' = \mathbf{A}' + \mathbf{B}'$$

$$\textbf{(Mat9)}\ (\mathbf{AB})' = \mathbf{B}'\mathbf{A}'$$

$$\textbf{(Mat10)}\ (\mathbf{A} - \mathbf{B})' = \mathbf{A}' - \mathbf{B}'$$

We will prove rules **Mat6–Mat8**.

THEOREM 6.1: *Let* **A** *be an* $m \times n$ *matrix. Then*

$$(\mathbf{A}')' = \mathbf{A}$$

Recall that, in order to prove that two matrices of the same order are equal, we need to show that they are equal elementwise. For simplicity, we will refer to the (i, j)th element of matrix **A** as $[A]_{ij}$. Then, for example, $\left[(A')'\right]_{ij}$ denotes the (i, j)th element of matrix $(\mathbf{A}')'$.

So, we need to show that

$$[(A')']_{ij} = [A]_{ij}$$

Proof: By definition of a transpose we have

$$[A]_{ij} = [A']_{ji} \tag{i}$$

and

$$[(A')']_{ij} = [A']_{ji} \tag{ii}$$

Combining (i) and (ii) we have

$$[A]_{ij} = [(A')']_{ij}$$

THEOREM 6.2: *Let k be a scalar and* **A** *be an* $m \times n$ *matrix. Then*

$$(k\mathbf{A})' = k\mathbf{A}'$$

Proof: By definition of a transpose and definition of scalar multiplication we have

$$[(kA)']_{ij} = [(kA)]_{ji} = k\,[A]_{ji} = k[A']_{ij} = [kA']_{ij}$$

Therefore,

$$(k\mathbf{A})' = k\mathbf{A}'$$

THEOREM 6.3: *Let* **A** *and* **B** *be two matrices of the same order* $(m \times n)$*. Then*

$$(\mathbf{A}+\mathbf{B})' = \mathbf{A}' + \mathbf{B}'$$

Proof: By definition of matrix transpose and definition of matrix addition:

$$[(A+B)']_{ij} = [(A+B)]_{ji} = [A]_{ji} + [B]_{ji} = [A']_{ij} + [B']_{ij}$$

Therefore,

$$(\mathbf{A}+\mathbf{B})' = \mathbf{A}' + \mathbf{B}'$$

Now, let us consider the following definition.

DEFINITION 6.1 Let **A** be a square matrix of order $n \times n$. If

$$\mathbf{A}' = \mathbf{A}$$

then **A** is called *symmetric*. Furthermore, if

$$\mathbf{A}' = -\mathbf{A}$$

then **A** is called *anti-symmetric*.

EXAMPLE 6.18 Let

$$\mathbf{A} = \begin{bmatrix} 3 & 4 & 7 \\ 4 & 2 & 11 \\ 7 & 11 & 1 \end{bmatrix}$$

then **A** is symmetric.

EXAMPLE 6.19 There are three coffee shops in a small college town. Each coffee shop sells four blends of coffee: Espresso, Cappuccino, Classic and French. The cost of coffee for a cup of each blend is listed in Table 6.1 and does not vary with the shop.

Table 6.1 The cost of coffee per cup

Espresso	\$1.50
Cappuccino	\$0.75
Classic	\$0.50
French	\$1.00

Table 6.2 Number of cups of coffee sold per week

	Espresso	*Cappuccino*	*Classic*	*French*
Coffee Shop 1	112	100	80	35
Coffee Shop 2	182	160	110	58
Coffee Shop 3	206	192	130	76

Table 6.3 Retail price (\$) of each blend at each shop

	Espresso	*Cappuccino*	*Classic*	*French*
Coffee Shop 1	8	4	3	6
Coffee Shop 2	6	2	2	5
Coffee Shop 3	5	3	1	3

The number of cups of coffee sold at each shop as well as retail prices are given in Tables 6.2 and 6.3.

Use matrix algebra to calculate:

(a) the total weekly cost of coffee for each shop,
(b) the total weekly revenue for each blend for each shop,
(c) the total weekly profit for each shop.

Solution:

(a) The number of cups sold can be written as

$$\mathbf{Q} = \begin{bmatrix} 112 & 100 & 80 & 35 \\ 182 & 160 & 110 & 58 \\ 206 & 192 & 130 & 76 \end{bmatrix}$$

The cost of each blend of coffee can be expressed as a column vector

$$\mathbf{C} = \begin{bmatrix} 1.5 \\ 0.75 \\ 0.5 \\ 1 \end{bmatrix}$$

Write TC in matrix form as

$$\text{TC} = \begin{bmatrix} \text{Shop 1} \\ \text{Shop 2} \\ \text{Shop 3} \end{bmatrix} = \mathbf{QC}$$

$$= \begin{bmatrix} 112 & 100 & 80 & 35 \\ 182 & 160 & 110 & 58 \\ 206 & 192 & 130 & 76 \end{bmatrix} \begin{bmatrix} 1.5 \\ 0.75 \\ 0.5 \\ 1 \end{bmatrix}$$

$$= \begin{bmatrix} 318 \\ 506 \\ 594 \end{bmatrix}$$

Thus, the respective costs are: Shop 1 = \$318, Shop 2 = \$506 and Shop 3 = \$594.

(b) By the same logic, we could write TR = **PQ**, where **P** is the price matrix

$$\mathbf{P} = \begin{bmatrix} 8 & 4 & 3 & 6 \\ 6 & 2 & 2 & 5 \\ 5 & 3 & 1 & 3 \end{bmatrix}$$

However, one can see immediately that it is not possible to multiply matrix $\mathbf{P}_{3\times4}$ by matrix $\mathbf{Q}_{3\times4}$. To overcome this problem, matrix **Q** must be transposed before multiplication. Therefore, we have

$$\mathbf{PQ}' = \begin{bmatrix} 8 & 4 & 3 & 6 \\ 6 & 2 & 2 & 5 \\ 5 & 3 & 1 & 3 \end{bmatrix} \begin{bmatrix} 112 & 182 & 206 \\ 100 & 160 & 192 \\ 80 & 110 & 130 \\ 35 & 58 & 76 \end{bmatrix}$$

$$= \begin{bmatrix} 1746 & 2774 & 3262 \\ 1207 & 1922 & 2260 \\ 1045 & 1674 & 1964 \end{bmatrix}$$

The total revenue for each shop is given by the elements of the main diagonal of the product matrix:

$$\text{TR} = \begin{bmatrix} \text{Shop 1} \\ \text{Shop 2} \\ \text{Shop 3} \end{bmatrix} = \begin{bmatrix} 1746 \\ 1922 \\ 1964 \end{bmatrix}$$

(c) Profit $\pi = \text{TR} - \text{TC}$ so

$$\pi = \begin{bmatrix} 1746 \\ 1922 \\ 1964 \end{bmatrix} - \begin{bmatrix} 318 \\ 506 \\ 594 \end{bmatrix} = \begin{bmatrix} 1428 \\ 1416 \\ 1370 \end{bmatrix}$$

Note that Shop 1 makes the highest profit.

PROBLEMS FOR SECTION 6.1

1. Let

$$\mathbf{a}=\begin{bmatrix}3\\5\\2\\1\end{bmatrix} \quad \text{and} \quad \mathbf{b}=\begin{bmatrix}3 & 1 & 2 & 4\end{bmatrix}$$

(a) Find **ab**.
(b) Find **ba**.

2. Let $k=5$,

$$\mathbf{A}=\begin{bmatrix}3 & 1 & 5\\7 & 5 & 2\\8 & 2 & -4\end{bmatrix} \quad \text{and} \quad \mathbf{B}=\begin{bmatrix}1 & -2 & 4\\9 & 5 & 7\\2 & 3 & -6\end{bmatrix}$$

(a) Find $\mathbf{A}+\mathbf{B}$, $k\mathbf{A}$, $k\mathbf{B}$, $(\mathbf{A}-\mathbf{B})$ and $(k\mathbf{A}+k\mathbf{B})$.
(b) Find $\mathbf{A}^2$, $\mathbf{B}^2$ and $(\mathbf{A}-\mathbf{B})^2$.
(c) Give an example of a matrix that is idempotent.
(d) Is product **AB** defined?
(e) Is product **BA** defined?

3. Let $k=8$,

$$\mathbf{A}=\begin{bmatrix}2 & 1\\-3 & 5\\4 & 8\end{bmatrix} \quad \text{and} \quad \mathbf{B}=\begin{bmatrix}4 & 1 & 5 & 1\\-5 & 8 & 2 & 3\end{bmatrix}$$

(a) Find **AB**, $k\mathbf{A}$ and $k\mathbf{B}$. Is the product **BA** defined?
(b) Find **IA**, **AI**, **IB** and **BI**. What is the dimension of matrix **I** in the product **IB**?

4. Let

$$\mathbf{A}=\begin{bmatrix}3 & 4 & 7\\-1 & 2 & 11\\12 & 14 & 1\end{bmatrix} \quad \text{and} \quad \mathbf{B}=\begin{bmatrix}1 & 3 & 4\\11 & -2 & 4\\1 & 3 & 2\end{bmatrix}$$

(a) Find $(\mathbf{A}-\mathbf{B})$ and $(\mathbf{A}+\mathbf{B})$.
(b) Show that $\mathbf{A}+\mathbf{B}=\mathbf{B}+\mathbf{A}$.

5. Show that, if matrix **A** is idempotent, then

(a) $(\mathbf{I}-\mathbf{A})$ is idempotent,
(b) $\mathbf{A}^k=\mathbf{A}$ for $k=5$.

6.2 Determinant of a matrix

For every square matrix **A** there is a scalar associated with it referred to as the determinant of that matrix and denoted as $|\mathbf{A}|$. The formal definition of the determinant of a square matrix is a difficult one and is beyond the scope of this book. Fortunately, there is a simple recursive procedure that allows us to calculate the determinant of any square matrix. According to the formal definition, the determinant of a 1×1 matrix

$$\mathbf{A} = [a_{11}]$$

is just the scalar a_{11},

$$|\mathbf{A}| = a_{11}$$

The determinant of a 1×1 matrix is called the *first-order determinant*. Now, the determinant of a 2×2 matrix

$$\mathbf{A} = \begin{bmatrix} a_{11} & a_{12} \\ a_{21} & a_{22} \end{bmatrix}$$

is given by

$$|\mathbf{A}| = a_{11}a_{22} - a_{12}a_{21}$$

The determinant of a 2×2 matrix is called the *second-order determinant*. In general, the determinant of an $n \times n$ matrix is called the *nth order determinant*.

EXAMPLE 6.20 Let

$$\mathbf{A} = \begin{bmatrix} 2 & 1 \\ 3 & 4 \end{bmatrix}$$

then

$$|\mathbf{A}| = 2 \times 4 - 1 \times 3 = 5$$

The following definition will help to define the determinant of an $n \times n$ matrix.

DEFINITION 6.2 Let $\mathbf{A}$ be a square matrix of order $n \times n$ and $\mathbf{A}_{ij}$ a submatrix obtained by deleting the ith row and jth column from $\mathbf{A}$. Let also

$$M_{ij} = |A_{ij}|$$
$$C_{ij} = (-1)^{i+j} M_{ij}$$

Then the scalar M_{ij} is called a *minor* of $\mathbf{A}$ and the scalar C_{ij} is called a *cofactor* of $\mathbf{A}$.

EXAMPLE 6.21 Let

$$\mathbf{A} = \begin{bmatrix} 7 & 1 & 4 \\ 3 & -2 & 1 \\ 5 & 11 & 23 \end{bmatrix}$$

then

$$M_{11} = \begin{vmatrix} -2 & 1 \\ 11 & 23 \end{vmatrix} = -57$$

$$M_{13} = \begin{vmatrix} 3 & -2 \\ 5 & 11 \end{vmatrix} = 43$$

$$M_{23} = \begin{vmatrix} 7 & 1 \\ 5 & 11 \end{vmatrix} = 72$$

$$C_{11} = (-1)^{1+1} M_{11} = -57$$

$$C_{13} = (-1)^{1+3} M_{13} = 43$$

$$C_{23} = (-1)^{2+3} M_{23} = -72$$

A method called Laplace expansion can be used to evaluate the determinant of a square matrix $\mathbf{A}_{n \times n}$. According to the Laplace expansion:

$$|\mathbf{A}| = \sum_{j=1}^{n} (a_{ij} C_{ij}) \quad \text{(expansion along the } i\text{th row)}$$

$$|\mathbf{A}| = \sum_{i=1}^{n} (a_{ij} C_{ij}) \quad \text{(expansion along the } j\text{th column)}$$

Example 6.22 Let $\mathbf{A}$ be a 3×3 matrix

$$\mathbf{A} = \begin{bmatrix} a_{11} & a_{12} & a_{13} \\ a_{21} & a_{22} & a_{23} \\ a_{31} & a_{32} & a_{33} \end{bmatrix}$$

Then expanding the determinant of $\mathbf{A}$ along the second row we have

$$\begin{aligned} \sum_{j=1}^{3} a_{2j} C_{2j} &= a_{21} C_{21} + a_{22} C_{22} + a_{23} C_{23} \\ &= a_{21}(-1)^{2+1} M_{21} + a_{22}(-1)^{2+2} M_{22} + a_{23}(-1)^{2+3} M_{23} \\ &= -a_{21} M_{21} + a_{22} M_{22} - a_{23} M_{23} \end{aligned}$$

Example 6.23 Let

$$\mathbf{A} = \begin{bmatrix} 1 & 0 & 2 \\ 3 & 2 & 5 \\ 7 & -4 & 11 \end{bmatrix}$$

Then expanding the determinant **A** along the first row we have

$$|\mathbf{A}| = (1)(-1)^{1+1}\begin{vmatrix} 2 & 5 \\ -4 & 11 \end{vmatrix} + (0)(-1)^{1+2}\begin{vmatrix} 3 & 5 \\ 7 & 11 \end{vmatrix} + (2)(-1)^{1+3}\begin{vmatrix} 3 & 2 \\ 7 & -4 \end{vmatrix}$$

$$= (1)\begin{vmatrix} 2 & 5 \\ -4 & 11 \end{vmatrix} + (2)\begin{vmatrix} 3 & 2 \\ 7 & -4 \end{vmatrix}$$

$$= (1)(42) + (2)(-26) = -10$$

Similarly, we can use Laplace expansion to evaluate determinants of order four and higher. For example, if matrix **A** has a dimension of 4 × 4, then

$$|\mathbf{A}| = a_{11}C_{11} + a_{12}C_{12} + a_{13}C_{13} + a_{14}C_{14}$$

where the cofactors are third-order determinants. Evaluation of higher-order determinants can also be reduced to evaluation of third-order determinants.

6.2.1 Properties of determinants

The following properties of determinants can be used to simplify the process of evaluating of a determinant.

PROPERTY 6.1 If a matrix has a row (column) of zeros, then the determinant is zero.

EXAMPLE 6.24

$$|\mathbf{A}| = \begin{vmatrix} 31 & 0 & -3 \\ 12 & 0 & 11 \\ 8 & 0 & 2 \end{vmatrix} = 0$$

This property follows immediately from the Laplace expansion (expand $|\mathbf{A}|$ by the second column).

PROPERTY 6.2 If a multiple of any row (column) is added to another row (column) of a matrix, then the determinant remains unchanged.

EXAMPLE 6.25

$$\begin{vmatrix} a_{11} & a_{12} \\ a_{21} + ka_{11} & a_{22} + ka_{12} \end{vmatrix} = a_{11}a_{22} + a_{11}ka_{12} - a_{12}a_{21} - a_{12}ka_{11}$$

$$= a_{11}a_{22} - a_{12}a_{21}$$

$$= \begin{vmatrix} a_{11} & a_{12} \\ a_{21} & a_{22} \end{vmatrix}$$

PROPERTY 6.3 A matrix with two identical rows (columns) has a determinant equal to zero.

Example 6.26

$$|\mathbf{A}| = \begin{vmatrix} 1 & 2 & 3 \\ 1 & 2 & 3 \\ 41 & 32 & 11 \end{vmatrix} = 0.$$

This property is an immediate consequence of Properties 1 and 2.

Property 6.4 A matrix that has two rows (columns) that are multiples of each other has a determinant that equals zero.

Example 6.27

$$|\mathbf{A}| = \begin{vmatrix} 1 & 2 & 3 \\ 5 & 10 & 15 \\ 41 & 32 & 11 \end{vmatrix} = 0$$

Here the second row is the multiple of the first one ($R2 = 5 \cdot R1$).

Property 6.5 Interchanging two rows (columns) of a matrix changes the sign, but not the absolute value, of the determinant.

Example 6.28

$$\begin{vmatrix} a_{21} & a_{22} \\ a_{11} & a_{12} \end{vmatrix} = a_{21}a_{12} - a_{22}a_{11} = -\begin{vmatrix} a_{11} & a_{12} \\ a_{21} & a_{22} \end{vmatrix}$$

Property 6.6 The determinant of the transpose of a matrix is equal to the determinant of the matrix ($|\mathbf{A}'| = |\mathbf{A}|$).

Example 6.29

$$\begin{vmatrix} a_{11} & a_{12} \\ a_{21} & a_{22} \end{vmatrix} = a_{11}a_{22} - a_{12}a_{21} = \begin{vmatrix} a_{11} & a_{21} \\ a_{12} & a_{22} \end{vmatrix}$$

Property 6.7 Multiplying a row (column) of a matrix $\mathbf{A}$ by a constant k multiplies the determinant by a factor of k.

Example 6.30

$$\begin{vmatrix} a_{11} & a_{12} \\ ka_{21} & ka_{22} \end{vmatrix} = k(a_{11}a_{22} - a_{12}a_{21}) = k\begin{vmatrix} a_{11} & a_{21} \\ a_{12} & a_{22} \end{vmatrix}$$

PROPERTY 6.8 If **A** is a triangular or diagonal matrix of order $n \times n$, then the determinant equals the product of the elements of the main diagonal:

$$|\mathbf{A}| = a_{11}a_{22}\ldots a_{nn}$$

EXAMPLE 6.31

$$\begin{vmatrix} 7 & 0 & 0 \\ 0 & -2 & 0 \\ 0 & 0 & 4 \end{vmatrix} = \begin{vmatrix} 7 & 0 & 0 \\ 23 & -2 & 0 \\ 14 & -3 & 4 \end{vmatrix} = \begin{vmatrix} 7 & 32 & 11 \\ 0 & -2 & 9 \\ 0 & 0 & 4 \end{vmatrix} = (7)(-2)(4) = -56$$

Determinants have many important applications, including solving systems of linear equations.

6.3 The matrix of cofactors

DEFINITION 6.3 Let **A** be a square matrix of order $n \times n$. Then

$$\mathbf{C} = \begin{bmatrix} C_{11} & C_{12} & \ldots & C_{1n} \\ C_{21} & C_{22} & \ldots & C_{nn} \\ \ldots & \ldots & \ldots & \ldots \\ C_{n1} & C_{n2} & \ldots & C_{nn} \end{bmatrix}$$

is called the *matrix of cofactors* of **A** and $\mathbf{C}'$ is called the *adjoint matrix* of **A** and written as $\text{Adj}(\mathbf{A}) = \mathbf{C}'$.

EXAMPLE 6.32 Let

$$\mathbf{A} = \begin{bmatrix} 1 & 2 & 3 \\ 4 & 1 & 2 \\ 3 & 5 & 4 \end{bmatrix}$$

Then the cofactor matrix is

$$\mathbf{C} = \begin{bmatrix} \begin{vmatrix} 1 & 2 \\ 5 & 4 \end{vmatrix} & -\begin{vmatrix} 4 & 2 \\ 3 & 4 \end{vmatrix} & \begin{vmatrix} 4 & 1 \\ 3 & 5 \end{vmatrix} \\ -\begin{vmatrix} 2 & 3 \\ 5 & 4 \end{vmatrix} & \begin{vmatrix} 1 & 3 \\ 3 & 4 \end{vmatrix} & -\begin{vmatrix} 1 & 2 \\ 3 & 5 \end{vmatrix} \\ \begin{vmatrix} 2 & 3 \\ 1 & 2 \end{vmatrix} & -\begin{vmatrix} 1 & 3 \\ 4 & 2 \end{vmatrix} & \begin{vmatrix} 1 & 2 \\ 4 & 1 \end{vmatrix} \end{bmatrix}$$

$$= \begin{bmatrix} -6 & -10 & 17 \\ 7 & -5 & 1 \\ 1 & 10 & -7 \end{bmatrix}$$

and the adjoint matrix is the transpose of the cofactor matrix,

$$\text{Adj}(\mathbf{A}) = \mathbf{C}' = \begin{bmatrix} -6 & 7 & 1 \\ -10 & -5 & 10 \\ 17 & 1 & -7 \end{bmatrix}$$

6.4 The inverse matrix

DEFINITION 6.4 Let $\mathbf{A}$ be a square matrix of order $n \times n$. If there exists a matrix $\mathbf{B}$ of order $n \times n$ such that

$$\mathbf{AB} = \mathbf{BA} = \mathbf{I}_n,$$

then $\mathbf{B}$ is called the *inverse matrix* of $\mathbf{A}$ (usually denoted as $\mathbf{B} = \mathbf{A}^{-1}$) and $\mathbf{A}$ is called *non-singular* (or invertible).

Furthermore, if $\mathbf{A}$ does not have an inverse, then $\mathbf{A}$ is called *singular*.

Note that it follows immediately from Definition 6.4 that, if $\mathbf{B}$ is the inverse of $\mathbf{A}$, then $\mathbf{A}$ must be the inverse of $\mathbf{B}$. In other words, $\mathbf{A}$ and $\mathbf{B}$ are inverses of each other.

PROPOSITION 6.1 *Let $\mathbf{A}$ be a square matrix of order $n \times n$. Then $\mathbf{A}$ is non-singular if and only if*

$$|\mathbf{A}| \neq 0$$

Moreover, if $\mathbf{A}$ is non-singular, then

$$\mathbf{A}^{-1} = \frac{1}{|\mathbf{A}|}\mathrm{Adj}(\mathbf{A})$$

and the inverse of $\mathbf{A}$ is unique.

EXAMPLE 6.33 Let

$$\mathbf{A} = \begin{bmatrix} 1 & 2 & 3 \\ 4 & 1 & 2 \\ 3 & 5 & 4 \end{bmatrix}$$

Then

$$\mathrm{Adj}(\mathbf{A}) = \begin{bmatrix} -6 & 7 & 1 \\ -10 & -5 & 10 \\ 17 & 1 & -7 \end{bmatrix}, \quad |\mathbf{A}| = 25$$

and

$$\mathbf{A}^{-1} = \frac{\mathrm{Adj}(\mathbf{A})}{|\mathbf{A}|} = \frac{1}{25}\begin{bmatrix} -6 & 7 & 1 \\ -10 & -5 & 10 \\ 17 & 1 & -7 \end{bmatrix}$$

$$= \begin{bmatrix} -0.24 & 0.28 & 0.04 \\ -0.4 & -0.2 & 0.4 \\ 0.68 & 0.04 & -0.28 \end{bmatrix}$$

6.5 Systems of linear equations

A system of m linear equations in n unknowns is represented by

$$\begin{cases} a_{11}x_1 + a_{12}x_2 + \cdots + a_{1n}x_n = b_1 \\ a_{21}x_1 + a_{22}x_2 + \cdots + a_{2n}x_n = b_2 \\ \qquad \cdots \\ a_{m1}x_1 + a_{m2}x_2 + \cdots + a_{mn}x_n = b_m \end{cases}$$

If

$$\mathbf{A} = \begin{bmatrix} a_{11} & a_{12} & \ldots & a_{1n} \\ a_{21} & a_{22} & \ldots & a_{2n} \\ \ldots & \ldots & \ldots & \ldots \\ a_{m1} & a_{m2} & \ldots & a_{mn} \end{bmatrix}, \quad \mathbf{x} = \begin{bmatrix} x_1 \\ x_2 \\ \ldots \\ x_n \end{bmatrix}, \quad \mathbf{b} = \begin{bmatrix} b_1 \\ b_2 \\ \ldots \\ b_m \end{bmatrix}$$

then the system above can be written in the following form:

$$\mathbf{Ax} = \mathbf{b}$$

A system of linear equations is called a *homogeneous system of linear equations* when

$$\mathbf{b} = \begin{bmatrix} b_1 \\ b_2 \\ \ldots \\ b_m \end{bmatrix} = \mathbf{0}$$

A system of linear equations is called a *non-homogeneous system of linear equations* when

$$\mathbf{b} \neq \mathbf{0}$$

Any vector x^* such that

$$\mathbf{Ax}^* = \mathbf{b}$$

is called a solution of the system of linear equations. If one of the solutions of the system of linear equations is $\mathbf{x}^* = \mathbf{0}$, the system is said to have a *trivial solution*. A solution of the system of linear equations that is different from the trivial one is called a *non-trivial solution*.

A system of linear equations is called *consistent* if it has at least one solution. A system of linear equations is called *inconsistent* if it has no solutions.

EXAMPLE 6.34 A homogeneous system of linear equations

$$\begin{cases} a_{11}x_1 + a_{12}x_2 + \cdots + a_{1n}x_n = 0 \\ a_{21}x_1 + a_{22}x_2 + \cdots + a_{2n}x_n = 0 \\ \qquad \cdots \\ a_{m1}x_1 + a_{m2}x_2 + \cdots + a_{mn}x_n = 0 \end{cases}$$

is always consistent since it has at least one solution: the trivial one. Note that a non-homogeneous system of linear equations does not have a trivial solution.

EXAMPLE 6.35 The system of non-homogeneous linear equations

$$\begin{cases} 2x_1 + 5x_2 = 5 \\ 2x_1 + 5x_2 = 7 \end{cases}$$

is inconsistent, since $(2x_1 + 5x_2)$ cannot be simultaneously equal to 5 and to 7.

6.5.1 Solving systems of linear equations with a non-zero determinant

In this section we will consider systems of linear equations

$$\mathbf{Ax} = \mathbf{b} \tag{6.1}$$

where $|\mathbf{A}| \neq 0$ and $m = n$.

If the determinant of $\mathbf{A}$ is not zero, then it follows that $\mathbf{A}$ is non-singular and we can premultiply each side of (6.1) by the inverse of $\mathbf{A}$:

$$\mathbf{A}^{-1}\mathbf{Ax} = \mathbf{A}^{-1}\mathbf{b} \tag{6.2}$$

Recall that by the definition of the inverse matrix

$$\mathbf{A}^{-1}\mathbf{A} = \mathbf{I} \tag{6.3}$$

Now, combining (6.2) and (6.3) we have

$$\mathbf{x} = \mathbf{A}^{-1}\mathbf{b}$$

PROPOSITION 6.2 *A system of linear equations with a non-zero determinant has a unique solution* $\mathbf{x}^*$ *given by*

$$\mathbf{x}^* = \mathbf{A}^{-1}\mathbf{b} \tag{6.4}$$

Moreover, if the system of equations is homogeneous ($\mathbf{b} = \mathbf{0}$), *then*

$$\mathbf{x}^* = \mathbf{0}$$

The method of solving a system of linear equations using formula (6.4) is called the *method of matrix inversion.*

EXAMPLE 6.36 Solve the following system of linear equations by matrix inversion.

$$\begin{cases} 3x + 4y = 7 \\ 4x + 5y = 8 \end{cases}$$

Solution: First, we express the system in a matrix form

$$\begin{bmatrix} 3 & 4 \\ 4 & 5 \end{bmatrix} \begin{bmatrix} x \\ y \end{bmatrix} = \begin{bmatrix} 7 \\ 8 \end{bmatrix}$$

Then, using equation (6.4) we have

$$\begin{bmatrix} x \\ y \end{bmatrix} = \begin{bmatrix} 3 & 4 \\ 4 & 5 \end{bmatrix}^{-1} \begin{bmatrix} 7 \\ 8 \end{bmatrix} = \begin{bmatrix} -3 \\ 4 \end{bmatrix}$$

There is yet another method of solving systems of linear equations called Cramer's rule. Given a system of linear equations in (6.1) Cramer's rule states that

$$x_i = \frac{|\mathbf{A}_i|}{|\mathbf{A}|} \tag{6.5}$$

where $\mathbf{A}_i$ is the matrix formed from the matrix of coefficients $\mathbf{A}$ by replacing the ith column of $\mathbf{A}$ with the column vector of constants $\mathbf{b}$.

EXAMPLE 6.37 Solve the following system of linear equations using Cramer's rule.

$$\begin{cases} 3x + 4y = 7 \\ 4x + 5y = 8 \end{cases}$$

Solution: First, we form matrices

$$\mathbf{A}_1 = \begin{bmatrix} 7 & 4 \\ 8 & 5 \end{bmatrix} \quad \text{and} \quad \mathbf{A}_2 = \begin{bmatrix} 3 & 7 \\ 4 & 8 \end{bmatrix}$$

from the original matrix of coefficients

$$\mathbf{A} = \begin{bmatrix} 3 & 4 \\ 4 & 5 \end{bmatrix}$$

Then, using equation (6.5) we have

$$x = \frac{\begin{vmatrix} 7 & 4 \\ 8 & 5 \end{vmatrix}}{\begin{vmatrix} 3 & 4 \\ 4 & 5 \end{vmatrix}} = -3$$

and

$$y = \frac{\begin{vmatrix} 3 & 7 \\ 4 & 8 \end{vmatrix}}{\begin{vmatrix} 3 & 4 \\ 4 & 5 \end{vmatrix}} = 4$$

EXAMPLE 6.38: (CONSUMPTION AND LEISURE MODEL). Consider an example of the consumer optimization problem of consumption C and leisure L where the representative consumer preferences (C and L) are perfect complements and

$$C = aL \tag{6.6}$$

where a is a constant with $a > 0$. Moreover, the quantities of consumption and leisure must satisfy the budget constraint:

$$C = w(h - L) + \pi - T \tag{6.7}$$

where h is a consumer's time endowment, w is the real wage, π represents real dividend income and T stands for taxes. Combining (6.6) and (6.7) we obtain

$$\begin{cases} C = aL \\ C = w(h - L) + \pi - T \end{cases}$$

Assuming that the only endogenous variables in the system are consumption and leisure, and rearranging the equations, gives

$$\begin{cases} C - aL = 0 \\ C + wL = wh + \pi - T \end{cases}$$

Expressing the system above in matrix form we have

$$\begin{bmatrix} 1 & -a \\ 1 & w \end{bmatrix} \times \begin{bmatrix} C \\ L \end{bmatrix} = \begin{bmatrix} 0 \\ wh + \pi - T \end{bmatrix}$$

Now, using Cramer's rule we have

$$C^* = \frac{\begin{vmatrix} 0 & -a \\ wh + \pi - T & w \end{vmatrix}}{\begin{vmatrix} 1 & -a \\ 1 & w \end{vmatrix}} = \frac{a(wh + \pi - T)}{a + w}$$

and

$$L^* = \frac{\begin{vmatrix} 1 & 0 \\ 1 & wh + \pi - T \end{vmatrix}}{\begin{vmatrix} 1 & -a \\ 1 & w \end{vmatrix}} = \frac{wh + \pi - T}{a + w}$$

Alternatively, we can solve the system using matrix inversion:

$$\begin{bmatrix} C^* \\ L^* \end{bmatrix} = \frac{1}{a + w} \times \begin{bmatrix} w & a \\ -1 & 1 \end{bmatrix} \begin{bmatrix} 0 \\ wh + \pi - T \end{bmatrix}$$

$$= \begin{bmatrix} \dfrac{a(wh + \pi - T)}{a + w} \\ \\ \dfrac{wh + \pi - T}{a + w} \end{bmatrix}$$

Note that both Cramer's rule and the matrix inversion method produce the same result.

PROBLEMS FOR SECTION 6.5

1. Find the following determinants.

(a) $|\mathbf{A}_1| = \begin{vmatrix} 3 & 0 \\ 12 & 7 \end{vmatrix}$

(b) $|\mathbf{A}_2| = \begin{vmatrix} 1 & 2 \\ 3 & 4 \end{vmatrix}$

(c) $|\mathbf{A}_3| = \begin{vmatrix} 3 & 2 & 1 \\ 4 & 3 & 2 \\ 5 & 4 & 3 \end{vmatrix}$

(d) $|\mathbf{A}_4| = \begin{vmatrix} 1 & 43 & 21 & 0 & 3 \\ 8 & 3 & 9 & 0 & 11 \\ 77 & 2 & 43 & 0 & 42 \\ 45 & 267 & 438 & 0 & 76 \\ 6 & 89 & -1 & 0 & 43 \end{vmatrix}$

(e) $|\mathbf{A}_5| = \begin{vmatrix} 1 & 32 & -11 & 31 & 6 \\ 0 & 2 & 12 & 67 & 9 \\ 0 & 0 & 3 & 33 & 4 \\ 0 & 0 & 0 & 2 & 6 \\ 0 & 0 & 0 & 0 & 5 \end{vmatrix}$

(f) $|\mathbf{A}_6| = \begin{vmatrix} 1 & 0 & 0 & 0 & 0 \\ 0 & 2 & 0 & 0 & 0 \\ 0 & 0 & 3 & 0 & 0 \\ 0 & 0 & 0 & 2 & 0 \\ 0 & 0 & 0 & 0 & 5 \end{vmatrix}$

(g) $|\mathbf{A}_7| = \begin{vmatrix} 1 & 2 & 3 & 4 & 5 \\ 1 & 2 & 3 & 4 & 5 \\ 321 & 19 & 11 & 32 & 15 \\ 61 & 15 & -21 & -5 & -2 \\ 35 & 34 & 62 & 59 & 51 \end{vmatrix}$

(h) $|\mathbf{A}_8| = |41|$

2. Find the transposes of the following matrices.

(a) $\mathbf{A}_1 = \begin{bmatrix} 3 & 2 & 1 \\ 4 & 3 & 2 \\ 5 & 4 & 3 \end{bmatrix}$

(b) $\mathbf{A}_2 = \begin{bmatrix} a & b & c \\ d & e & f \\ g & h & k \end{bmatrix}$

(c) $\mathbf{A}_3 = \begin{bmatrix} 43 & 12 & 3 & 1 & 6 \\ 11 & -3 & 34 & 76 & 3 \\ 65 & 26 & 47 & 51 & 18 \end{bmatrix}$

3. Find the inverses of the following matrices.

(a) $\mathbf{A}_1 = \begin{bmatrix} 3 & 1 \\ 2 & 4 \end{bmatrix}$

(b) $\mathbf{A}_2 = \begin{bmatrix} 1 & 0 & 0 \\ 0 & 1 & 0 \\ 0 & 0 & 1 \end{bmatrix}$

4. Solve the following systems of linear equations by matrix inversion and using Cramer's rule.

(a) $\begin{cases} 2x + 3y = 5 \\ 5x - 2y = 8 \end{cases}$

(b) $\begin{cases} x + y + z = 6 \\ 2x - y + 3z = 9 \\ 3x + 4y - z = 9 \end{cases}$

5. Let

$$\mathbf{A} = \begin{bmatrix} 1 & 0 & 0 \\ 0 & 1 & 0 \\ 0 & 0 & 1 \end{bmatrix}, \quad \mathbf{B} = \begin{bmatrix} 2 & 3 \\ 1 & 5 \\ 3 & 7 \end{bmatrix}, \quad \mathbf{C} = \begin{bmatrix} 1 & 2 \\ 3 & 4 \end{bmatrix}$$

(a) Find $\mathbf{A}'$ and $\mathbf{B}'$.
(b) Show that $(\mathbf{ABC})' = \mathbf{C}'\mathbf{B}'\mathbf{A}'$.

6. Consider the following simple macroeconomic model:

$$\begin{cases} Y = C + I_0 + G_0 \\ C = a + bY \end{cases}$$

where a and b are constants with $a > 0$ and $0 < b < 1$, Y is national income, C is consumption, I_0 is investment and G_0 is government expenditures. The endogenous variables in this system are Y and C. Find Y and C using Cramer's rule and matrix inversion.

7 Further topics in matrix algebra

An important occurrence of matrices comes from linearizing dependences. When certain quantities depend on several quantities, and not just one variable, the corresponding rates of change naturally form a matrix. Mathematical properties of this matrix reveal a lot about the dependence. When maximizing functions of one variable, we often looked at the second derivative. In a situation when a function depends on several variables, the second derivatives would form a matrix, called the Hessian. This construction, and related bordered Hessian matrices, play an important role in optimization theory.

This chapter sees a significant increase in computational difficulty. While smaller systems of equations may be easily solved by hand, using, for instance, the substitution method, as the dimension of systems increases it becomes necessary to use specialized software. In this book we use Mathematica from Wolfram Research. This computer algebra system (CAS) has seen significant improvements in recent years. Among others, a significantly better numerical engine was introduced that now approaches the computational speed of rival software such as Matlab.

Our application of Mathematica will be focused on issues of linear dependence and rank. These notions are crucial in figuring out whether a given system of linear equations has a unique solution.

7.1 Linear dependence

DEFINITION 7.1 A set of k vectors $\{\mathbf{v}_1, \mathbf{v}_2, \ldots, \mathbf{v}_k\}$ of the same dimension is *linearly independent* if the equation

$$\sum_{i=1}^{k} c_i \mathbf{v}_i = \mathbf{0} \tag{7.1}$$

where the c_i are scalars, holds only if $c_1 = c_2 = \cdots = c_k = 0$. Otherwise, the vectors are *linearly dependent*.

Without loss of generality, let us assume that the vectors in question are all $m \times 1$ column vectors. Now we can put equation (7.1) in a matrix form:

$$\mathbf{Ac} = \sum_{i=1}^{k} c_i \mathbf{v}_i = \mathbf{0}, \quad \text{where } \mathbf{A} = [\mathbf{v}_1\ \mathbf{v}_2\ \ldots\ \mathbf{v}_k] \text{ and } \mathbf{c} = \begin{bmatrix} c_1 \\ c_2 \\ \vdots \\ c_k \end{bmatrix} \tag{7.2}$$

Now we see that the set of k vectors $\{\mathbf{v}_1, \mathbf{v}_2, \ldots, \mathbf{v}_k\}$ is linearly independent if and only if the homogeneous system $\mathbf{Ac} = \mathbf{0}$ has a non-trivial solution.

EXAMPLE 7.1 Determine whether the vectors below are linearly dependent or independent.

$$\mathbf{v}_1 = \begin{bmatrix} 1 \\ 2 \\ 3 \end{bmatrix}, \quad \mathbf{v}_2 = \begin{bmatrix} 2 \\ 4 \\ 6 \end{bmatrix}, \quad \mathbf{v}_1 = \begin{bmatrix} 3 \\ 6 \\ 9 \end{bmatrix}$$

Solution: First, we form the associated homogeneous system:

$$\mathbf{Ac} = \begin{bmatrix} 1 & 2 & 3 \\ 2 & 4 & 6 \\ 3 & 6 & 9 \end{bmatrix} \begin{bmatrix} c_1 \\ c_2 \\ c_3 \end{bmatrix} = \begin{bmatrix} 0 \\ 0 \\ 0 \end{bmatrix}$$

Multiply the first row of $\mathbf{A}$ by 6, the second row of $\mathbf{A}$ by 3 and the third row of $\mathbf{A}$ by 2 to get

$$\begin{bmatrix} 6 & 12 & 18 \\ 6 & 12 & 18 \\ 6 & 12 & 18 \end{bmatrix} \begin{bmatrix} c_1 \\ c_2 \\ c_3 \end{bmatrix} = \begin{bmatrix} 0 \\ 0 \\ 0 \end{bmatrix} \tag{7.3}$$

Note that the homogeneous system reduces to one equation with three unknowns:

$$6c_1 + 12c_2 + 18c_3 = 0 \tag{7.4}$$

Observe that equation (7.4) has infinitely many solutions. If we set $c_1 = c_2 = 1$, then it follows readily that $c_3 = -1$. This shows that the homogeneous system in (7.2) has a non-trivial solution. Thus, the set of vectors $\{\mathbf{v}_1, \mathbf{v}_2, \mathbf{v}_3\}$ is linearly dependent.

PROPOSITION 7.1 *A set of k column vectors $\{\mathbf{v}_1, \mathbf{v}_2, \ldots, \mathbf{v}_k\}$ is linearly independent if and only if the associated matrix $\mathbf{A} = [\mathbf{v}_1, \mathbf{v}_2, \ldots, \mathbf{v}_k]$ has rank k.*

DEFINITION 7.2 The *rank* of a matrix $\mathbf{A}$ is the number of linearly independent rows of $\mathbf{A}$.

Since the number of linearly independent rows is equal to the number of linearly independent columns, we can can say that the rank of a matrix is the number of its linearly independent rows or columns.

7.2 Quadratic forms

A quadratic form in n variables is a polynomial of the form

$$Q(x_1, x_2, \ldots, x_n) = \sum_{i=1}^{n} \sum_{j=1}^{n} a_{ij} x_i x_j$$

where $a_{11}, a_{12}, \ldots, a_{nn}$ are some real numbers, known as the coefficients of the quadratic form.

A quadratic form can be written as

$$Q(x_1, x_2, \ldots, x_n) = \mathbf{x}'\mathbf{A}\mathbf{x}$$

where $\mathbf{A}$ is a symmetric matrix and $\mathbf{x}' = [x_1\ x_2\ \ldots x_n]$.

EXAMPLE 7.2 A quadratic form in two variables is

$$\begin{aligned} Q(x_1, x_2) &= \sum_{i=1}^{2}\sum_{j=1}^{2} a_{ij}x_i x_j \\ &= a_{11}x_1^2 + a_{12}x_1x_2 + a_{21}x_2x_1 + a_{22}x_2^2 \end{aligned}$$

EXAMPLE 7.3 A quadratic form in three variables is

$$Q(x_1, x_2, x_3) = 3x_1^2 + 5x_2^2 - 7x_3^2 + x_1x_2 + 2x_1x_3$$

EXAMPLE 7.4 Consider the following quadratic form:

$$Q(x, y) = 2x^2 + 6xy + y^2$$

Find its associated symmetric matrix.

Solution: To find its associated symmetric matrix

$$\mathbf{A} = \begin{bmatrix} a_{11} & a_{12} \\ a_{21} & a_{22} \end{bmatrix}$$

we write $Q(x, y)$ in the following way:

$$\begin{aligned} Q(x, y) &= \begin{bmatrix} x & y \end{bmatrix} \begin{bmatrix} a_{11} & a_{12} \\ a_{21} & a_{22} \end{bmatrix} \begin{bmatrix} x \\ y \end{bmatrix} \\ &= \begin{bmatrix} x & y \end{bmatrix} \begin{bmatrix} a_{11} & a_{12} \\ a_{12} & a_{22} \end{bmatrix} \begin{bmatrix} x \\ y \end{bmatrix} \\ &= a_{11}x^2 + 2a_{12}xy + a_{22}y^2 \end{aligned}$$

It follows immediately that

$$Q(x, y) = \begin{bmatrix} x & y \end{bmatrix} \begin{bmatrix} 2 & 3 \\ 3 & 1 \end{bmatrix} \begin{bmatrix} x \\ y \end{bmatrix}$$

DEFINITION 7.3 A quadratic form $Q(x_1, x_2, \ldots, x_n)$ and its associated symmetric matrix is called:

positive definite	if $Q > 0\ \forall \mathbf{x} \neq \mathbf{0}$,
negative definite	if $Q < 0\ \forall \mathbf{x} \neq \mathbf{0}$,
positive semidefinite	if $Q \geq 0\ \forall \mathbf{x} \neq \mathbf{0}$,
negative semidefinite	if $Q \leq 0\ \forall \mathbf{x} \neq \mathbf{0}$.

A quadratic form $Q(x_1, x_2, \ldots, x_n)$ and its associated symmetric matrix is called *indefinite* if it is neither positive semidefinite nor negative semidefinite.

EXAMPLE 7.5

(a) A quadratic form $Q(x, y, z) = x^2 + y^2 + z^2$ is positive definite since $Q > 0$ always for all $(x, y, z) \neq (0, 0, 0)$.
(b) A quadratic form $Q(x, y, z) = (x + y - z)^2$ is positive semidefinite since $Q \geq 0$ always, and $Q = 0$ for $z = x + y$.
(c) A quadratic form $Q(x, y, z) = -(x + y - z)^2$ is negative semidefinite since $Q \leq 0$ always, and $Q = 0$ for $z = x + y$.
(d) A quadratic form $Q(x, y, z) = x^2 + y^2 - z^2$ is indefinite, since $Q(1, 1, 1) > 0$ and $Q(1, 1, 5) < 0$.

EXAMPLE 7.6 Show that the identity matrix, **I**, is positive definite.

Solution: The identity matrix, **I**, is positive definite since

$$\mathbf{x}'\mathbf{I}\mathbf{x} = \mathbf{x}'\mathbf{x} > \mathbf{0} \quad \text{if } \mathbf{x} \neq \mathbf{0}$$

DEFINITION 7.4 Let **B** be a square matrix of dimension n. We define Ω_k, the kth-order leading principal minor of **B**, as

$$\Omega_1 = |b_{11}|$$

$$\Omega_2 = \begin{vmatrix} b_{11} & b_{12} \\ b_{21} & b_{22} \end{vmatrix}$$

and in general

$$\Omega_k = \begin{vmatrix} b_{11} & b_{12} & \cdots & b_{1k} \\ b_{21} & b_{22} & \cdots & b_{2k} \\ \vdots & \vdots & \ddots & \vdots \\ b_{k1} & b_{k2} & \cdots & b_{kk} \end{vmatrix}$$

We also define an arbitrary principal minor of order k as the determinant of a matrix obtained by deleting all but k rows and columns of **B** with the same numbers.

Note that matrix **B** has exactly n leading principal minors.

EXAMPLE 7.7 Find the leading principal minors of the following matrices:

$$\mathbf{A} = \begin{bmatrix} a_{11} & a_{12} & a_{13} \\ a_{21} & a_{22} & a_{23} \\ a_{31} & a_{32} & a_{33} \end{bmatrix} \quad \text{and} \quad \mathbf{B} = \begin{bmatrix} b_{11} & b_{12} & b_{13} & b_{14} \\ b_{21} & b_{22} & b_{23} & b_{24} \\ b_{31} & b_{32} & b_{33} & b_{34} \\ b_{41} & b_{42} & b_{43} & b_{44} \end{bmatrix}$$

Solution: The leading principal minors of **A** are

$$|a_{11}|, \quad \begin{vmatrix} a_{11} & a_{12} \\ a_{21} & a_{22} \end{vmatrix}, \quad \begin{vmatrix} a_{11} & a_{12} & a_{13} \\ a_{21} & a_{22} & a_{23} \\ a_{31} & a_{32} & a_{33} \end{vmatrix}$$

The leading principal minors of **B** are

$$|b_{11}|, \quad \begin{vmatrix} b_{11} & b_{12} \\ b_{21} & b_{22} \end{vmatrix}, \quad \begin{vmatrix} b_{11} & b_{12} & b_{13} \\ b_{21} & b_{22} & b_{23} \\ b_{31} & b_{32} & b_{33} \end{vmatrix}, \quad \begin{vmatrix} b_{11} & b_{12} & b_{13} & b_{14} \\ b_{21} & b_{22} & b_{23} & b_{24} \\ b_{31} & b_{32} & b_{33} & b_{34} \\ b_{41} & b_{42} & b_{43} & b_{44} \end{vmatrix}$$

EXAMPLE 7.8 Find all the principal minors of the following matrices:

$$\mathbf{A} = \begin{bmatrix} a_{11} & a_{12} & a_{13} \\ a_{21} & a_{22} & a_{23} \\ a_{31} & a_{32} & a_{33} \end{bmatrix} \quad \text{and} \quad \mathbf{B} = \begin{bmatrix} b_{11} & b_{12} & b_{13} & b_{14} \\ b_{21} & b_{22} & b_{23} & b_{24} \\ b_{31} & b_{32} & b_{33} & b_{34} \\ b_{41} & b_{42} & b_{43} & b_{44} \end{bmatrix}$$

Solution: The principal minors of **A** are

$$|a_{11}|, \quad |a_{22}|, \quad |a_{33}|, \quad \begin{vmatrix} a_{11} & a_{12} \\ a_{21} & a_{22} \end{vmatrix}, \quad \begin{vmatrix} a_{11} & a_{13} \\ a_{31} & a_{33} \end{vmatrix}, \quad \begin{vmatrix} a_{22} & a_{23} \\ a_{32} & a_{33} \end{vmatrix}, \quad |\mathbf{A}|.$$

The principal minors of **B** of order one are

$$|b_{11}|, \quad |b_{22}|, \quad |b_{33}|, \quad |b_{44}|$$

The principal minors of **B** of order two are

$$\begin{vmatrix} b_{33} & b_{34} \\ b_{43} & b_{44} \end{vmatrix}, \quad \begin{vmatrix} b_{22} & b_{24} \\ b_{42} & b_{44} \end{vmatrix}, \quad \begin{vmatrix} b_{22} & b_{23} \\ b_{32} & b_{33} \end{vmatrix},$$

$$\begin{vmatrix} b_{11} & b_{14} \\ b_{21} & b_{24} \end{vmatrix}, \quad \begin{vmatrix} b_{11} & b_{13} \\ b_{31} & b_{33} \end{vmatrix}, \quad \begin{vmatrix} b_{11} & b_{12} \\ b_{21} & b_{22} \end{vmatrix}$$

The principal minors of **B** of order three are

$$\begin{vmatrix} b_{22} & b_{23} & b_{24} \\ b_{32} & b_{33} & b_{34} \\ b_{42} & b_{43} & b_{44} \end{vmatrix}, \quad \begin{vmatrix} b_{11} & b_{13} & b_{14} \\ b_{31} & b_{33} & b_{34} \\ b_{41} & b_{43} & b_{44} \end{vmatrix}, \quad \begin{vmatrix} b_{11} & b_{12} & b_{14} \\ b_{21} & b_{22} & b_{24} \\ b_{41} & b_{42} & b_{44} \end{vmatrix}, \quad \begin{vmatrix} b_{11} & b_{12} & b_{13} \\ b_{21} & b_{22} & b_{23} \\ b_{31} & b_{32} & b_{33} \end{vmatrix}$$

And the principal minor of **B** of order four is

$$|\mathbf{B}|$$

Note that a 4×4 matrix **B** has a total of 15 principal minors and only four leading principal minors.

THEOREM 7.1: *A symmetric matrix* **A** *is:*

positive definite	$\Longleftrightarrow$	*all of its leading principal minors are positive,*
negative definite	$\Longleftrightarrow$	*its leading principal minors alternate in sign with the first one being negative,*
positive semidefinite	$\Longleftrightarrow$	*all of its principal minors are non-negative,*
negative semidefinite	$\Longleftrightarrow$	$(-1)^{k+1}\Omega_k \leq 0\ \forall\, k = 1, 2, \ldots, n.$

A symmetric matrix **A** *is* indefinite *if it is neither positive semidefinite nor negative semidefinite.*

PROBLEMS FOR SECTION 7.2

1. Determine whether the following matrices are positive definite, negative definite, positive semidefinite, negative semidefinite or indefinite.

(a) $\mathbf{A}_1 = \begin{bmatrix} 1 & 2 \\ 2 & 5 \end{bmatrix}$

(b) $\mathbf{A}_2 = \begin{bmatrix} -1 & 1 \\ -22 & 5 \end{bmatrix}$

(c) $\mathbf{A}_3 = \begin{bmatrix} -7 & 12 \\ -2 & 1 \end{bmatrix}$

(d) $\mathbf{A}_4 = \begin{bmatrix} 3 & 5 \\ 2 & 18 \end{bmatrix}$

2. Determine whether the following matrices are positive definite, negative definite, positive semidefinite, negative semidefinite or indefinite.

(a) $\mathbf{B}_1 = \begin{bmatrix} 1 & 0 & 0 \\ 0 & 1 & 0 \\ 0 & 0 & 1 \end{bmatrix}, \quad \mathbf{B}_2 = \begin{bmatrix} 1 & 1 & 1 \\ 1 & 1 & 1 \\ 1 & 1 & 1 \end{bmatrix}, \quad \mathbf{B}_3 = \begin{bmatrix} 1 & 2 & 3 \\ 2 & 1 & 2 \\ 3 & 2 & 1 \end{bmatrix}$

(b) $\mathbf{C}_1 = \begin{bmatrix} 1 & 1 & 1 & 1 \\ 1 & 1 & 1 & 1 \\ 1 & 1 & 1 & 1 \\ 1 & 1 & 1 & 1 \end{bmatrix}, \quad \mathbf{C}_2 = \begin{bmatrix} 1 & 0 & 0 & 0 \\ 0 & 1 & 0 & 0 \\ 0 & 0 & 1 & 0 \\ 0 & 0 & 0 & 1 \end{bmatrix}, \quad \mathbf{C}_3 = \begin{bmatrix} 1 & 2 & 3 & 4 \\ 5 & 6 & 7 & 8 \\ 9 & 10 & 11 & 12 \\ 13 & 14 & 15 & 16 \end{bmatrix}$

7.3 The Hessian matrix

Let $y = f(x_1, x_2, \ldots, x_n)$ be a scalar-valued function of n variables. If $\mathbf{x} = (x_1, x_2, \ldots, x_n)$, then

$$y = f(\mathbf{x})$$

and the *vector of partial derivatives* (also known as the *gradient vector*) is given by

$$\frac{\partial f(\mathbf{x})}{\partial \mathbf{x}} = \begin{bmatrix} f_1 \\ f_2 \\ \vdots \\ f_n \end{bmatrix} = \begin{bmatrix} \frac{\partial y}{\partial x_1} \\ \frac{\partial y}{\partial x_2} \\ \vdots \\ \frac{\partial y}{\partial x_n} \end{bmatrix}$$

The dimension of the gradient vector is $n \times 1$. We can think of the gradient vector as the matrix of first derivarives. The *Hessian matrix* (also known as the *Hessian* or the *second-derivatives matrix*) of $f(x_1, x_2, \ldots, x_n)$ is computed as

$$\mathbf{H} = \begin{bmatrix} f_{11} & f_{12} & \cdots & f_{1n} \\ f_{21} & f_{22} & \cdots & f_{2n} \\ \vdots & \vdots & \ddots & \vdots \\ f_{n1} & f_{n2} & \cdots & f_{nn} \end{bmatrix}$$

The *Hessian determinant* of $f(x_1, x_2, \ldots, x_n)$ is

$$|\mathbf{H}| = \begin{vmatrix} f_{11} & f_{12} & \cdots & f_{1n} \\ f_{21} & f_{22} & \cdots & f_{2n} \\ \vdots & \vdots & \ddots & \vdots \\ f_{n1} & f_{n2} & \cdots & f_{nn} \end{vmatrix}$$

and it plays a very important role in optimization problems.

The leading principal minors of the Hessian matrix are

$$|\mathbf{H}_1| = |f_{11}|$$

$$|\mathbf{H}_2| = \begin{vmatrix} f_{11} & f_{12} \\ f_{21} & f_{22} \end{vmatrix}$$

$$|\mathbf{H}_3| = \begin{vmatrix} f_{11} & f_{12} & f_{13} \\ f_{21} & f_{22} & f_{23} \\ f_{31} & f_{32} & f_{33} \end{vmatrix}$$

and

$$|\mathbf{H}_n| = |\mathbf{H}| = \begin{vmatrix} f_{11} & f_{12} & \cdots & f_{1n} \\ f_{21} & f_{22} & \cdots & f_{2n} \\ \vdots & \vdots & \ddots & \vdots \\ f_{n1} & f_{n2} & \cdots & f_{nn} \end{vmatrix}$$

7.4 Row echelon form of a matrix

In matrix algebra, a matrix is in a *row echelon form* if:

(a) all non-zero rows are above any other rows of all zeros,
(b) the pivot (the first non-zero number from the left) of a non-zero row is always strictly to the right of the pivot of the row above it.

A matrix is in a *reduced row echelon* form if it is in a row echelon form and every pivot is 1 and is the only non-zero entry in its column.

EXAMPLE 7.9 The identity matrix

$$\mathbf{I}_4 = \begin{bmatrix} 1 & 0 & 0 & 0 \\ 0 & 1 & 0 & 0 \\ 0 & 0 & 1 & 0 \\ 0 & 0 & 0 & 1 \end{bmatrix}$$

is in a row echelon form. Note that the identity matrix has only non-zero rows. In addition, the identity matrix is in a reduced row echelon form.

EXAMPLE 7.10 The zero matrix below is in a row echelon form.

$$\mathbf{0}_{4\times 5} = \begin{bmatrix} 0 & 0 & 0 & 0 & 0 \\ 0 & 0 & 0 & 0 & 0 \\ 0 & 0 & 0 & 0 & 0 \\ 0 & 0 & 0 & 0 & 0 \end{bmatrix}$$

EXAMPLE 7.11 The following matrices are in row echelon form.

$$\mathbf{A}_1 = \begin{bmatrix} 0 & 0 & 2 & 1 & 3 & 4 & 7 & 2 \\ 0 & 0 & 0 & 7 & 0 & 0 & 0 & 0 \\ 0 & 0 & 0 & 0 & 2 & 1 & 2 & 5 \end{bmatrix}$$

$$\mathbf{A}_2 = \begin{bmatrix} 0 & 7 & 0 & 3 & 2 \\ 0 & 0 & 0 & 1 & 0 \\ 0 & 0 & 0 & 0 & 0 \\ 0 & 0 & 0 & 0 & 0 \end{bmatrix}$$

$$\mathbf{A}_3 = \begin{bmatrix} 1 & 7 & 0 & 0 \\ 0 & 17 & 0 & 0 \\ 0 & 0 & 17 & 0 \\ 0 & 0 & 0 & 17 \end{bmatrix}$$

7.5 The rank of a matrix

Let $\mathbf{A}$ be a matrix of dimension $m \times n$. It could be shown that the number of linearly independent rows of $\mathbf{A}$ is equal to the number of independent columns of $\mathbf{A}$. The number of independent rows of $\mathbf{A}$ is known as the rank of $\mathbf{A}$, and denoted Rank($\mathbf{A}$). It follows readily that

$$\text{Rank}(\mathbf{A}) \leq \min(m, n)$$

To compute the rank of a matrix we need to perform elementary row operations until the matrix is transformed to a row echelon form. The number of non-zero rows remaining in the row echelon form is going to be the rank of the matrix.

EXAMPLE 7.12 Find the rank of the matrix

$$\mathbf{A} = \begin{bmatrix} 1 & 2 & 3 \\ 4 & 5 & 6 \\ 7 & 8 & 9 \end{bmatrix}$$

Solution: Note that the rank of $\mathbf{A}$ is less than or equal to 3. Now, let us perform the following row equivalent operations:

$$\begin{bmatrix} 1 & 2 & 3 \\ 4 & 5 & 6 \\ 7 & 8 & 9 \end{bmatrix} \xrightarrow{R_2 - 4R_1} \begin{bmatrix} 1 & 2 & 3 \\ 0 & -3 & -6 \\ 7 & 8 & 9 \end{bmatrix}$$

$$\xrightarrow{R_2 * \left(-\frac{1}{3}\right)} \begin{bmatrix} 1 & 2 & 3 \\ 0 & 1 & 2 \\ 7 & 8 & 9 \end{bmatrix} \xrightarrow{R_1 - R_2} \begin{bmatrix} 1 & 1 & 1 \\ 0 & 1 & 2 \\ 7 & 8 & 9 \end{bmatrix}$$

$$\xrightarrow{R_3 - R_2} \begin{bmatrix} 1 & 1 & 1 \\ 0 & 1 & 2 \\ 7 & 7 & 7 \end{bmatrix} \xrightarrow{R_3 - 7R_1} \begin{bmatrix} 1 & 1 & 1 \\ 0 & 1 & 2 \\ 0 & 0 & 0 \end{bmatrix}$$

Observe that the matrix

$$\mathbf{B} = \begin{bmatrix} 1 & 1 & 1 \\ 0 & 1 & 2 \\ 0 & 0 & 0 \end{bmatrix}$$

is in row echelon form and has two non-zero rows. We conclude that Rank($\mathbf{A}$) $= 2$.

DEFINITION 7.5 Two matrices of the same dimension are *row equivalent* if one can be obtained from the other by means of elementary row operations. Elementary row operations include interchanging two rows, multiplying a row by a non-zero number, or adding a row to another row multiplied by a number.

For instance, the two matrices below are row equivalent:

$$\mathbf{A} = \begin{bmatrix} 1 & 1 & 1 \\ 0 & 1 & 2 \\ 7 & 7 & 7 \end{bmatrix} \quad \text{and} \quad \mathbf{B} = \begin{bmatrix} 1 & 1 & 1 \\ 0 & 1 & 2 \\ 0 & 0 & 0 \end{bmatrix}$$

Elementary row operations can be used in solving linear systems and finding the inverse.

DEFINITION 7.6 A matrix $\mathbf{A}_{m\times n}$ is called *full rank* if

$$\text{Rank}(\mathbf{A}_{m\times n}) = \min(m, n)$$

For example, the identity matrix

$$\mathbf{I}_3 = \begin{bmatrix} 1 & 0 & 0 \\ 0 & 1 & 0 \\ 0 & 0 & 1 \end{bmatrix}$$

is full rank.

THEOREM 7.2: *Suppose the system* $\mathbf{Ax} = \mathbf{b}$ *has m equations and n variables. We denote the augmented matrix of the system by* $(\mathbf{A}/\mathbf{b})$. *Then we have that:*

(i) *the system* $\mathbf{Ax} = \mathbf{b}$ *is consistent if and only if* $\text{Rank}(\mathbf{A}) = \text{Rank}(\mathbf{A}/\mathbf{b})$
(ii) *the system* $\mathbf{Ax} = \mathbf{b}$ *has a unique solution if and only if*

$$\text{Rank}(\mathbf{A}) = \text{Rank}(\mathbf{A}/\mathbf{b}) = n$$

(iii) *if* $\text{Rank}(\mathbf{A}) < n$ *there are infinitely many solutions.*

To understand the concept of augmented matrix, consider the system

$$\begin{bmatrix} 2 & 3 \\ 4 & 5 \end{bmatrix} \begin{bmatrix} x_1 \\ x_2 \end{bmatrix} = \begin{bmatrix} 7 \\ 8 \end{bmatrix}$$

Then the augmented matrix of the system is defined as

$$(\mathbf{A}/\mathbf{b}) = \begin{bmatrix} 2 & 3 & 7 \\ 4 & 5 & 8 \end{bmatrix}$$

PROBLEMS FOR SECTION 7.5

1. Determine whether the following matrices are full rank.

(a) $\mathbf{A}_1 = \begin{bmatrix} 1 & 2 & 3 & 4 \\ 1 & 2 & 5 & 6 \\ 1 & 2 & 7 & 8 \end{bmatrix}$

(b) $\mathbf{A}_2 = \begin{bmatrix} 1 & 2 & 3 & 4 \\ 5 & 6 & 7 & 8 \\ 9 & 10 & 11 & 12 \\ 13 & 14 & 15 & 16 \end{bmatrix}$

2. Find the reduced row echelon form and rank of the following matrices.

(a) $\mathbf{A}_1 = \begin{bmatrix} 1 & 2 & 3 \\ 1 & 2 & 4 \\ 1 & 2 & 5 \end{bmatrix}$

(b) $\mathbf{A}_2 = \begin{bmatrix} 1 & 2 & 3 \\ 1 & 2 & 5 \\ 1 & 2 & 7 \end{bmatrix}$

(c) $\mathbf{A}_3 = \begin{bmatrix} 1 & 2 & 3 \\ 1 & 2 & 5 \\ 1 & 2 & 7 \end{bmatrix}$

(d) $\mathbf{A}_4 = \begin{bmatrix} 1 & 2 & 3 & 4 \\ 5 & 6 & 7 & 8 \\ 9 & 10 & 11 & 12 \\ 13 & 14 & 15 & 16 \end{bmatrix}$

3. Determine whether the following matrices are full rank.

(a) $\mathbf{A}_1 = \begin{bmatrix} 1 & 2 & 3 \\ 1 & 2 & 5 \\ 1 & 2 & 7 \end{bmatrix}$

(b) $\mathbf{A}_2 = \begin{bmatrix} 1 & 2 \\ -5 & 6 \\ 9 & 10 \\ 13 & 14 \end{bmatrix}$

7.6 Eigenvalues and eigenvectors

Definition 7.7 Let **A** be a square matrix of order n. If λ is a scalar and **x** is a non-zero vector such that

$$\mathbf{Ax} = \lambda \mathbf{x} \tag{7.5}$$

then **x** is called the *eigenvector* of **A**. Further, the scalar λ is called an *eigenvalue* of matrix **A** corresponding to the eigenvector **x**.

A square matrix can have both real and complex eigenvalues. The equation above can be written as

$$(\mathbf{A} - \lambda \mathbf{I})\mathbf{x} = \mathbf{0} \tag{7.6}$$

which is a homogeneous system with n equations and n unknowns. Recall that a homogeneous system with n equations and n unknowns has a non-trivial solution only if the determinant of the coefficient matrix equals zero:

$$|(\mathbf{A} - \lambda \mathbf{I})| = \begin{vmatrix} (a_{11} - \lambda) & a_{12} & \dots & a_{1n} \\ a_{21} & (a_{22} - \lambda) & \dots & a_{2n} \\ \dots & \dots & \dots & \dots \\ a_{n1} & a_{n2} & \dots & (a_{nn} - \lambda) \end{vmatrix} = 0 \tag{7.7}$$

The equation above is called the eigenvalue equation of matrix **A**. The eigenvalue equation has exactly n solutions $(\lambda_1, \lambda_2, \ldots, \lambda_n)$, which may or may not be distinct. If

$$\mathbf{x} = \begin{bmatrix} x_1 \\ x_2 \\ \ldots \\ x_n \end{bmatrix}$$

then equation (7.6) can be written as

$$\begin{bmatrix} (a_{11}-\lambda) & a_{12} & \ldots & a_{1n} \\ a_{21} & (a_{22}-\lambda) & \ldots & a_{2n} \\ \ldots & \ldots & \ldots & \ldots \\ a_{n1} & a_{n2} & \ldots & (a_{nn}-\lambda) \end{bmatrix} \begin{bmatrix} x_1 \\ x_2 \\ \ldots \\ x_n \end{bmatrix} = \mathbf{0}$$

or alternatively as

$$\begin{cases} (a_{11}-\lambda)x_1 + a_{12}x_2 + \cdots + a_{1n}x_n = 0 \\ a_{21}x_1 + (a_{22}-\lambda)x_2 + \cdots + a_{2n}x_n = 0 \\ \qquad \ldots \\ a_{n1}x_1 + a_{n2}x_2 + \cdots + (a_{nn}-\lambda)x_n = 0 \end{cases} \tag{7.8}$$

The following examples will shed some light on the theory above.

EXAMPLE 7.13 Find the eigenvalues and the associated eigenvectors of these matrices.

$$\mathbf{A} = \begin{bmatrix} 1 & 4 \\ 3 & -3 \end{bmatrix}, \quad \mathbf{B} = \begin{bmatrix} 1 & 0 \\ 7 & 2 \end{bmatrix}, \quad \mathbf{C} = \begin{bmatrix} 3 & 4 \\ 5 & 2 \end{bmatrix}$$

Solution: The eigenvalue equation of **A** is given by

$$|\mathbf{A} - \lambda\mathbf{I}| = \begin{vmatrix} 1-\lambda & 4 \\ 3 & -3-\lambda \end{vmatrix} = \lambda^2 + 2\lambda - 15 = 0$$

The equation above has two solutions, $\lambda_1 = -5$ and $\lambda_2 = 3$. Therefore, $\lambda_1 = -5$ and $\lambda_2 = 3$ are the eigenvalues of **A**.

Substituting $\lambda_1 = -5$ in system (7.8) we have

$$\begin{cases} (1+5)x_1 + 4x_2 = 0 \\ 3x_1 + (-3+5)x_2 = 0 \end{cases}$$

The system above reduces to

$$3x_1 + 2x_2 = 0$$

Choosing $x_2 = k$, we have $x_1 = -\frac{2}{3}k$. The eigenvectors associated with $\lambda_1 = -5$ are $x = k\binom{-\frac{2}{3}}{1}$, $k \neq 0$. We substitute $k = -3q$ to equivalently represent the eigenvectors associated with $\lambda_1 = -5$ as $x = q\binom{-2}{3}$, $q \neq 0$.

For $\lambda_2 = 3$, system (7.8) implies that

$$x_1 - 2x_2 = 0$$

Choosing $x_2 = k$, we have $x_1 = 2k$. The eigenvectors associated with $\lambda_1 = 3$ are $x = k\binom{2}{1}$, $k \neq 0$.

The eigenvalue equation of $\mathbf{B}$ is

$$|\mathbf{B} - \lambda\mathbf{I}| = \begin{vmatrix} 1-\lambda & 0 \\ 7 & 2-\lambda \end{vmatrix} = \lambda^2 - 3\lambda + 2 = 0$$

This has solutions $\lambda_1 = 2$ and $\lambda_2 = 1$, which are the eigenvalues of $\mathbf{B}$.

Substituting $\lambda_1 = 2$ in system (7.8) we have

$$\begin{cases} (1-2)x_1 + 0x_2 = 0 \\ 7x_1 + (2-2)x_2 = 0 \end{cases}$$

The system above reduces to

$$x_1 = 0$$

Choosing $x_2 = k$, we have $x_1 = 0$. The eigenvectors associated with $\lambda_1 = 2$ are $x = k\binom{0}{1}$, $k \neq 0$.

For $\lambda_2 = 1$, system (7.8) reduces to

$$7x_1 + x_2 = 0.$$

Choosing $x_2 = 7k$, we have $x_1 = -k$. The eigenvectors associated with $\lambda_1 = 1$ are $x = k\binom{-1}{7}$, $k \neq 0$.

The eigenvalue equation of $\mathbf{C}$ is

$$|\mathbf{C} - \lambda\mathbf{I}| = \begin{vmatrix} 3-\lambda & 4 \\ 5 & 2-\lambda \end{vmatrix} = \lambda^2 - 5\lambda - 14 = 0$$

The equation above has solutions $\lambda_1 = 7$ and $\lambda_2 = -2$, which are the eigenvalues of $\mathbf{C}$.

Substituting $\lambda_1 = 7$ in system (7.8) we have

$$\begin{cases} (3-7)x_1 + 4x_2 = 0 \\ 5x_1 + (2-7)x_2 = 0 \end{cases}$$

The system above reduces to

$$-x_1 + x_2 = 0$$

Choosing $x_2 = k$, we have $x_1 = k$. The eigenvectors associated with $\lambda_1 = 7$ are $x = k\binom{1}{1}$, $k \neq 0$.

For $\lambda_2 = -2$, system (7.8) implies

$$5x_1 + 4x_2 = 0$$

Choosing $x_2 = 5k$, we have $x_1 = -4k$. Hence, the eigenvectors associated with $\lambda_1 = -2$ are $x = k\binom{-4}{5}$, $k \neq 0$.

Example 7.14 Find the eigenvalues and eigenvectors of the matrix

$$\mathbf{A} = \begin{bmatrix} 2 & 2 & 2 \\ 1 & 5 & 0 \\ 2 & 2 & 2 \end{bmatrix}$$

Solution: The characteristic equation of **A** is

$$|\mathbf{A} - \lambda\mathbf{I}| = \begin{vmatrix} 2-\lambda & 2 & 2 \\ 1 & 5-\lambda & 0 \\ 2 & 2 & 2-\lambda \end{vmatrix} = \lambda^2(\lambda - 3)(\lambda - 6) = 0$$

The solutions of the equation above are

$$\lambda_1 = 6, \quad \lambda_2 = 3, \quad \lambda_3 = 0$$

Those lambdas are the distinct eigenvalues of matrix **A**. Let us now find the eigenvectors associated with λ_1, λ_2 and λ_3.

Substituting $\lambda_1 = 6$ in system (7.8) yields

$$\begin{bmatrix} (2-6)x_1 + 2x_2 + 2x_3 \\ x_1 + (5-6)x_2 \\ 2x_1 + 2x_2 + (2-6)x_3 \end{bmatrix} = \begin{bmatrix} 0 \\ 0 \\ 0 \end{bmatrix}$$

The solution to the system of linear equations above is $x_1 = x_2 = x_3$. Letting $x_3 = q$, we have $x_1 = x_2 = x_3 = q$. Therefore, the eigenvectors associated with $\lambda_1 = 6$ are given by

$$\mathbf{x} = q \begin{bmatrix} 1 \\ 1 \\ 1 \end{bmatrix}$$

where $q \neq 0$.

Substituting $\lambda_2 = 3$ in system (7.8) yields

$$\begin{bmatrix} (2-3)x_1 + 2x_2 + 2x_3 \\ x_1 + (5-3)x_2 \\ 2x_1 + 2x_2 + (2-3)x_3 \end{bmatrix} = \begin{bmatrix} 0 \\ 0 \\ 0 \end{bmatrix}$$

The solution to the system of linear equations above is $x_1 = x_3$ and $x_2 = -\frac{1}{2}x_1$. Letting $x_1 = q$, we have $x_1 = x_3 = q$ and $x_2 = -\frac{1}{2}q$. Therefore, the eigenvectors associated with $\lambda_2 = 3$ are given by

$$\mathbf{x} = q \begin{bmatrix} -5 \\ 1 \\ 4 \end{bmatrix}$$

Substituting $\lambda_3 = 0$ in system (7.8) yields

$$\begin{bmatrix} 2x_1 + 2x_2 + 2x_3 \\ x_1 + 5x_2 \\ 2x_1 + 2x_2 + 2x_3 \end{bmatrix} = \begin{bmatrix} 0 \\ 0 \\ 0 \end{bmatrix}$$

The solution to the system of linear equations above is $x_1 = -5x_2$ and $x_3 = 4x_2$. Letting $x_2 = q$, we have

$$\begin{cases} x_1 = -5q \\ x_2 = q \\ x_3 = 4q \end{cases}$$

Therefore, the eigenvectors associated with $\lambda_1 = 0$ are given by

$$\mathbf{x} = q \begin{bmatrix} -5 \\ 1 \\ 4 \end{bmatrix}$$

7.7 Kronecker product

The Kronecker product (named after German mathematician Leopold Kronecker) is a matrix operation transforming two matrices into a block matrix.

Consider two matrices with different dimensions, $\mathbf{A} = [a_{ij}]$ and $\mathbf{B} = [b_{ij}]$. Let $\text{Dim}(\mathbf{A}) = m \times n$ and $\text{Dim}(\mathbf{B}) = p \times q$. The Kronecker product of $\mathbf{A}$ and $\mathbf{B}$ is defined as

$$\mathbf{A} \otimes \mathbf{B} = \begin{bmatrix} a_{11}\mathbf{B} & a_{12}\mathbf{B} & \dots & a_{1n}\mathbf{B} \\ a_{21}\mathbf{B} & a_{22}\mathbf{B} & \dots & a_{2n}\mathbf{B} \\ \vdots & \vdots & \ddots & \vdots \\ a_{m1}\mathbf{B} & a_{m2}\mathbf{B} & \dots & a_{mn}\mathbf{B} \end{bmatrix}$$

and

$$\mathbf{B} \otimes \mathbf{A} = \begin{bmatrix} b_{11}\mathbf{A} & b_{12}\mathbf{A} & \dots & b_{1q}\mathbf{A} \\ b_{21}\mathbf{A} & b_{22}\mathbf{A} & \dots & b_{2q}\mathbf{A} \\ \vdots & \vdots & \ddots & \vdots \\ b_{p1}\mathbf{A} & b_{p2}\mathbf{A} & \dots & b_{pq}\mathbf{A} \end{bmatrix}$$

EXAMPLE 7.15 Let

$$\mathbf{B} = \begin{bmatrix} 1 & 2 & 3 \\ 2 & 2 & 2 \end{bmatrix} \quad \text{and} \quad \mathbf{D} = \begin{bmatrix} 1 & 2 \\ 3 & 4 \end{bmatrix}$$

Then, their Kronecker product is

$$\mathbf{B}\otimes\mathbf{D}=\begin{bmatrix} 1\cdot\begin{bmatrix}1 & 2\\ 3 & 4\end{bmatrix} & 2\cdot\begin{bmatrix}1 & 2\\ 3 & 4\end{bmatrix} & 3\cdot\begin{bmatrix}1 & 2\\ 3 & 4\end{bmatrix} \\ 2\cdot\begin{bmatrix}1 & 2\\ 3 & 4\end{bmatrix} & 2\cdot\begin{bmatrix}1 & 2\\ 3 & 4\end{bmatrix} & 2\cdot\begin{bmatrix}1 & 2\\ 3 & 4\end{bmatrix} \end{bmatrix}$$

$$=\begin{bmatrix} \begin{bmatrix}1 & 2\\ 3 & 4\end{bmatrix} & \begin{bmatrix}2 & 4\\ 6 & 8\end{bmatrix} & \begin{bmatrix}3 & 6\\ 9 & 12\end{bmatrix} \\ \begin{bmatrix}2 & 4\\ 6 & 8\end{bmatrix} & \begin{bmatrix}2 & 4\\ 6 & 8\end{bmatrix} & \begin{bmatrix}2 & 4\\ 6 & 8\end{bmatrix} \end{bmatrix}$$

EXAMPLE 7.16 Let

$$\mathbf{A}=\begin{bmatrix}a_{11} & a_{12} & a_{13}\\ a_{21} & a_{22} & a_{23}\end{bmatrix} \quad \text{and} \quad \mathbf{K}=\begin{bmatrix}1 & 2\\ 2 & 1\\ 1 & 2\end{bmatrix}$$

Then, their Kronecker product is

$$\mathbf{A}\otimes\mathbf{K}=\begin{bmatrix} a_{11}\begin{bmatrix}1 & 2\\ 2 & 1\\ 1 & 2\end{bmatrix} & a_{12}\begin{bmatrix}1 & 2\\ 2 & 1\\ 1 & 2\end{bmatrix} & a_{13}\begin{bmatrix}1 & 2\\ 2 & 1\\ 1 & 2\end{bmatrix} \\ a_{21}\begin{bmatrix}1 & 2\\ 2 & 1\\ 1 & 2\end{bmatrix} & a_{22}\begin{bmatrix}1 & 2\\ 2 & 1\\ 1 & 2\end{bmatrix} & a_{23}\begin{bmatrix}1 & 2\\ 2 & 1\\ 1 & 2\end{bmatrix} \end{bmatrix}$$

$$=\begin{bmatrix} \begin{bmatrix}a_{11} & 2a_{11}\\ 2a_{11} & a_{11}\\ a_{11} & 2a_{11}\end{bmatrix} & \begin{bmatrix}a_{12} & 2a_{12}\\ 2a_{12} & a_{12}\\ a_{12} & 2a_{12}\end{bmatrix} & \begin{bmatrix}a_{13} & 2a_{13}\\ 2a_{13} & a_{13}\\ a_{13} & 2a_{13}\end{bmatrix} \\ \begin{bmatrix}a_{21} & 2a_{21}\\ 2a_{21} & a_{21}\\ a_{21} & 2a_{21}\end{bmatrix} & \begin{bmatrix}a_{22} & 2a_{22}\\ 2a_{22} & a_{22}\\ a_{22} & 2a_{22}\end{bmatrix} & \begin{bmatrix}a_{23} & 2a_{23}\\ 2a_{23} & a_{23}\\ a_{23} & 2a_{23}\end{bmatrix} \end{bmatrix}$$

EXAMPLE 7.17 Let

$$\mathbf{I}=\begin{bmatrix}i_{11} & i_{12}\\ i_{21} & i_{22}\end{bmatrix} \quad \text{and} \quad \mathbf{F}=\begin{bmatrix}f_{11} & f_{12}\\ f_{21} & f_{22}\end{bmatrix}$$

The Kronecker product of **I** and **F** will be

$$\mathbf{I}\otimes\mathbf{F}=\begin{bmatrix} i_{11}\begin{bmatrix} f_{11} & f_{12}\\ f_{21} & f_{22}\end{bmatrix} & i_{12}\begin{bmatrix} f_{11} & f_{12}\\ f_{21} & f_{22}\end{bmatrix}\\ i_{21}\begin{bmatrix} f_{11} & f_{12}\\ f_{21} & f_{22}\end{bmatrix} & i_{22}\begin{bmatrix} f_{11} & f_{12}\\ f_{21} & f_{22}\end{bmatrix}\end{bmatrix}$$

$$=\begin{bmatrix} \begin{bmatrix} i_{11}f_{11} & i_{11}f_{12}\\ i_{11}f_{21} & i_{11}f_{22}\end{bmatrix} & \begin{bmatrix} i_{12}f_{11} & i_{12}f_{12}\\ i_{12}f_{21} & i_{12}f_{22}\end{bmatrix}\\ \begin{bmatrix} i_{21}f_{11} & i_{21}f_{12}\\ i_{21}f_{21} & i_{21}f_{22}\end{bmatrix} & \begin{bmatrix} i_{22}f_{11} & i_{22}f_{12}\\ i_{22}f_{21} & i_{22}f_{22}\end{bmatrix}\end{bmatrix}$$

Note the following six properties of Kronecker products:

(Kro1) $(\mathbf{A}\otimes\mathbf{B})'=\mathbf{A}'\otimes\mathbf{B}'$
(Kro2) $(\mathbf{A}\otimes\mathbf{B})(\mathbf{C}\otimes\mathbf{D})=\mathbf{AC}\otimes\mathbf{BD}$
(Kro3) $\mathbf{A}\otimes(\mathbf{B}+\mathbf{C})=\mathbf{A}\otimes\mathbf{B}+\mathbf{A}\otimes\mathbf{C}$
(Kro4) $(\mathbf{B}+\mathbf{C})\otimes\mathbf{A}=\mathbf{B}\otimes\mathbf{A}+\mathbf{C}\otimes\mathbf{A}$
(Kro5) $\mathbf{A}\otimes(\mathbf{B}\otimes\mathbf{C})=(\mathbf{A}\otimes\mathbf{B})\otimes\mathbf{C}$
(Kro6) $(\mathbf{A}\otimes\mathbf{B})^{-1}=\mathbf{A}^{-1}\otimes\mathbf{B}^{-1}$

7.8 Vectorization of a matrix

Matrix vectorization is a matrix operation that transforms a matrix into a column vector by placing the columns on the top of each other (with the first column to be on the top and the last column to be on the bottom).

Let us vectorize matrix $\mathbf{A}=[a_{ij}]$ with dimensions $q\times p$. Then

$$\text{Vec}(\mathbf{A})=\begin{bmatrix} a_{11}\\ \vdots\\ a_{q1}\\ a_{12}\\ \vdots\\ a_{q2}\\ \vdots\\ a_{1p}\\ \vdots\\ a_{qp}\end{bmatrix}$$

EXAMPLE 7.18 Let

$$\mathbf{I}=\begin{bmatrix} 10 & 20\\ 30 & 40\end{bmatrix}\quad\text{and}\quad\mathbf{F}=\begin{bmatrix} 5 & -5 & 15\\ 2 & 11 & -3\end{bmatrix}$$

Then

$$\text{Vec}(\mathbf{I}) = \begin{bmatrix} 10 \\ 30 \\ 20 \\ 40 \end{bmatrix} \quad \text{and} \quad \text{Vec}(\mathbf{F}) = \begin{bmatrix} 5 \\ 2 \\ -5 \\ 11 \\ 15 \\ -3 \end{bmatrix}$$

Example 7.19 Let

$$\mathbf{O} = \begin{bmatrix} o_{11} & o_{12} \\ o_{21} & o_{22} \end{bmatrix} \quad \text{and} \quad \mathbf{P} = \begin{bmatrix} 1 & 4 & 0 \\ 2 & 0 & -13 \\ 0 & 32 & 44 \end{bmatrix}$$

Then

$$\text{Vec}(\mathbf{O}) = \begin{bmatrix} o_{11} \\ o_{21} \\ o_{12} \\ o_{22} \end{bmatrix} \quad \text{and} \quad \text{Vec}(\mathbf{P}) = \begin{bmatrix} 1 \\ 2 \\ 0 \\ 4 \\ 0 \\ 32 \\ 0 \\ -13 \\ 44 \end{bmatrix}$$

Example 7.20 Let

$$\mathbf{Y} = \begin{bmatrix} y_{11} & y_{12} \\ y_{21} & y_{22} \\ y_{31} & y_{32} \end{bmatrix} \quad \text{and} \quad \mathbf{Z} = \begin{bmatrix} z_{11} & z_{12} & z_{13} & z_{14} \\ z_{21} & z_{22} & z_{23} & z_{24} \end{bmatrix}$$

Then

$$\text{Vec}(\mathbf{Y}) = \begin{bmatrix} y_{11} \\ y_{21} \\ y_{31} \\ y_{12} \\ y_{22} \\ y_{32} \end{bmatrix} \quad \text{and} \quad \text{Vec}(\mathbf{Z}) = \begin{bmatrix} z_{11} \\ z_{21} \\ z_{12} \\ z_{22} \\ z_{13} \\ z_{23} \\ z_{14} \\ z_{24} \end{bmatrix}$$

PROBLEMS FOR SECTION 7.8

1. Given the matrices

$$\mathbf{A} = \begin{bmatrix} 12 & 2 \\ 3 & 1 \end{bmatrix}, \quad \mathbf{B} = \begin{bmatrix} 2 & 4 & 5 \\ -5 & 12 & 1 \\ 2 & 11 & 3 \end{bmatrix}$$

find

(a) $\mathbf{A} \otimes \mathbf{B}$,
(b) $\mathbf{B} \otimes \mathbf{A}$.

2. Given the matrices

$$\mathbf{K} = \begin{bmatrix} 15 & 5 & -5 \\ 12 & 0 & -2 \end{bmatrix}, \quad \mathbf{L} = \begin{bmatrix} 1 & 5 & 3 & 4 \\ 0 & 2 & 9 & 8 \\ 3 & 1 & 6 & 9 \end{bmatrix}$$

$$\mathbf{M} = \begin{bmatrix} 28 & 2 & 3 \\ 27 & -5 & -10 \\ 26 & 12 & 4 \end{bmatrix}, \quad \mathbf{N} = \begin{bmatrix} 15 & 5 & -5 \\ 12 & 0 & -2 \end{bmatrix}$$

find

(a) $\mathbf{K} \otimes \mathbf{M}$,
(b) $\mathbf{K} \otimes \mathbf{L}$,
(c) $\mathbf{K} \otimes \mathbf{L} \otimes \mathbf{N}$,
(d) $\text{Vec}(\mathbf{K} \otimes \mathbf{N})$,
(e) $\text{Vec}(\mathbf{K} \otimes \mathbf{L} \otimes \mathbf{N})$.

3. Given the matrices

$$\mathbf{A} = \begin{bmatrix} 0 & 1 & 0 \\ -1 & 0 & -1 \\ 0 & 1 & 0 \end{bmatrix}, \quad \mathbf{B} = \begin{bmatrix} 2 & -3 & 4 \\ 1 & -2 & 5 \end{bmatrix}$$

verify the relationship in **Kro1.**

4. (a) Compute the eigenvalues and the eigenvectors of the following.

$$\mathbf{A}_1 = \begin{bmatrix} 1 & 1 \\ 1 & 1 \end{bmatrix}, \quad \mathbf{A}_2 = \begin{bmatrix} 1 & 0 \\ 0 & 1 \end{bmatrix}, \quad \mathbf{A}_3 = \begin{bmatrix} 2 & 2 \\ 2 & 2 \end{bmatrix}, \quad \mathbf{A}_4 = \begin{bmatrix} 2 & -5 \\ 3 & -6 \end{bmatrix}$$

(b) Compute the eigenvalues and the eigenvectors of the following.

$$\mathbf{B}_1 = \begin{bmatrix} 1 & 4 & 5 \\ 0 & -3 & 0 \\ 0 & 0 & -6 \end{bmatrix}, \quad \mathbf{B}_2 = \begin{bmatrix} 1 & 0 & 0 \\ 0 & 2 & 0 \\ 0 & 0 & 3 \end{bmatrix}, \quad \mathbf{B}_3 = \begin{bmatrix} 1 & 2 & 3 \\ 1 & 1 & 1 \\ 3 & 2 & 1 \end{bmatrix}$$

(c) Find all the eigenvalues and eigenvectors of

$$\mathbf{C} = \begin{bmatrix} 1 & 0 & 1 & 0 \\ 0 & 2 & 0 & 2 \\ 3 & 0 & 3 & 0 \\ 0 & 4 & 0 & 4 \end{bmatrix}$$

5. Given the matrices

$$\mathbf{A} = \begin{bmatrix} 0 & 1 \\ -1 & 0 \end{bmatrix}, \quad \mathbf{B} = \begin{bmatrix} 2 & -3 & 4 \\ 1 & -2 & 5 \end{bmatrix}, \quad \mathbf{C} = \begin{bmatrix} 2 & 3 \\ 4 & 7 \end{bmatrix}$$

verify the relationship in **Kro5.**

6. Given the matrices

$$\mathbf{C} = \begin{bmatrix} 2 & 1 & 0 \\ -2 & 0 & 1 \\ 0 & -1 & 3 \end{bmatrix}, \quad \mathbf{D} = \begin{bmatrix} 3 & 5 & 14 \\ 10 & -2 & 15 \end{bmatrix}, \quad \mathbf{M} = \begin{bmatrix} 2 & 4 & -8 \\ 3 & 5 & 9 \\ 11 & -3 & 10 \end{bmatrix}$$

verify the relationships in **Kro2** and **Kro4.**

7.9 Mathematica examples

EXAMPLE 7.21 Consider the matrix

$$\mathbf{I}_4 = \begin{bmatrix} 1 & 0 & 0 & 0 \\ 0 & 1 & 0 & 0 \\ 0 & 0 & 1 & 0 \\ 0 & 0 & 0 & 1 \end{bmatrix}$$

as in Example 7.9. Using the following code

```
> MatrixRank[{{1, 0, 0, 0}, {0, 1, 0, 0}, {0, 0, 1, 0}, {0, 0, 0, 1}}]
```

we find that the rank of the matrix $\mathbf{I}_4$ is equal to 4. Note that the matrix is in a row echelon form.

EXAMPLE 7.22 Consider the zero matrix below

$$\mathbf{0}_{4\times 5} = \begin{bmatrix} 0 & 0 & 0 & 0 & 0 \\ 0 & 0 & 0 & 0 & 0 \\ 0 & 0 & 0 & 0 & 0 \\ 0 & 0 & 0 & 0 & 0 \end{bmatrix}$$

Since the number of rows of the matrix is four, we know that the rank of the matrix does not exceed four. We use the code below

```
> MatrixRank[{{0, 0, 0, 0, 0}, {0, 0, 0, 0, 0}, {0, 0, 0, 0, 0}, {0, 0, 0, 0, 0}}]
```

to show that the rank of the matrix $\mathbf{0}_{4\times 5}$ is equal to 0. Since any two columns of the matrix are linearly dependent, the maximum number of linearly independent columns equals one. However, by convention, the rank of a zero matrix is equal to 0.

EXAMPLE 7.23 Find the rank of the following matrices.

$$\mathbf{A}_1 = \begin{bmatrix} 0 & 0 & 2 & 1 & 3 & 4 & 7 & 2 \\ 0 & 0 & 0 & 7 & 0 & 0 & 0 & 0 \\ 0 & 0 & 0 & 0 & 2 & 1 & 2 & 5 \end{bmatrix}$$

$$\mathbf{A}_2 = \begin{bmatrix} 0 & 7 & 0 & 3 & 2 \\ 0 & 0 & 0 & 1 & 0 \\ 0 & 0 & 0 & 0 & 0 \\ 0 & 0 & 0 & 0 & 0 \end{bmatrix}$$

$$\mathbf{A}_3 = \begin{bmatrix} 1 & 7 & 0 & 0 \\ 0 & 17 & 0 & 0 \\ 0 & 0 & 17 & 0 \\ 0 & 0 & 0 & 17 \end{bmatrix}$$

Solution: Using the following code

```
> MatrixRank[{{0, 0, 2, 1, 3, 4, 7, 2}, {0, 0, 0, 7, 0, 0, 0, 0}, {0, 0, 0, 0, 2, 1, 2, 5}}]
> MatrixRank[{{0, 7, 0, 3, 2}, {0, 0, 0, 1, 0}, {0, 0, 0, 0, 0}, {0, 0, 0, 0, 0}}]
> MatrixRank[{{1, 7, 0, 0}, {0, 17, 0, 0}, {0, 0, 17, 0}, {0, 0, 0, 17}}]
```

we find that

$$\text{Rank}(\mathbf{A}_1) = 3$$
$$\text{Rank}(\mathbf{A}_2) = 2$$
$$\text{Rank}(\mathbf{A}_3) = 4$$

Note that matrices $\mathbf{A}_1$ and $\mathbf{A}_3$ are full rank.

EXAMPLE 7.24 Solve the following linear system of equations:

$$\begin{cases} 3x - 2y + z = 1 \\ 5x - y + 3z = 14 \\ x - 4y - 2z = -19 \end{cases}$$

Solution: We can rewrite the system in a matrix form:

$$\begin{bmatrix} 3 & -2 & 1 \\ 5 & -1 & 3 \\ 1 & -4 & -2 \end{bmatrix} \begin{bmatrix} x \\ y \\ z \end{bmatrix} = \begin{bmatrix} 1 \\ 14 \\ -19 \end{bmatrix}$$

First, we find the rank of the coefficient matrix and augmented matrix using the following code:

> **MatrixRank[{{3, −2, 1}, {5, −1, 3}, {1, −4, −2}}]**

> **MatrixRank[{{3, −2, 1, 1}, {5, −1, 3, 14}, {1, −4, −2, −19}}]**

It follows that the rank of the coefficient matrix equals the rank of the augmented matrix:

$$\text{Rank}(\mathbf{A}) = \text{Rank}(\mathbf{A}/\mathbf{b}) = 3$$

So by Theorem 7.1, there is a unique solution to the system of linear equations. The code below allows us to find the solution:

> **Ainv = Inverse[{{3, −2, 1}, {5, −1, 3}, {1, −4, −2}}]**

> **b = {{1}, {14}, {−19}}**

> **Ainv.b**

It produces the following output:

```
{{-(14/3),8/3,5/3},{-(13/3),7/3,4/3},{19/3,-(10/3),-(7/3)}}
{{1},{14},{-19}}
{{1},{3},{4}}
```

So, the solution is

$$\begin{bmatrix} x \\ y \\ z \end{bmatrix} = \begin{bmatrix} 1 \\ 3 \\ 4 \end{bmatrix}$$

EXAMPLE 7.25 Find all solutions of the following system of linear equations:

$$\begin{cases} 2x - 3y + 2z - 5w = -21 \\ x + 5y - 4z + w = 0 \\ -3x + y - \frac{1}{4}z + 2w = 8 \\ 5x - 5y - z + 3w = 6 \end{cases}$$

Solution: First, we write the system in a matrix form as

$$\begin{bmatrix} 2 & -3 & 2 & -5 \\ 1 & 5 & -4 & 1 \\ -3 & 1 & -\frac{1}{4} & 2 \\ 5 & -5 & -1 & 3 \end{bmatrix} \begin{bmatrix} x \\ y \\ z \\ w \end{bmatrix} = \begin{bmatrix} -21 \\ 0 \\ 8 \\ 6 \end{bmatrix}$$

and then we find that the rank of the coefficient matrix equals the rank of the augmented matrix using the code

> **MatrixRank**[{{2, −3, 2, −5}, {1, 5, −4, 1}, {−3, 1, −1/4, 2}, {5, −5, −1, 3}}]
> **MatrixRank**[{{2, −3, 2, −5, −21}, {1, 5, −4, 1, 0}, {−3, 1, −1/4, 2, 8}, {5, −5, −1, 3, 6}}]

It follows that the rank of the coefficient matrix equals the rank of the augmented matrix:

$$\text{Rank}(\mathbf{A}) = \text{Rank}(\mathbf{A}/\mathbf{b}) = 4$$

There is a unique solution to the system of linear equations. The code below helps us to find the solution:

> **Ainv = Inverse**[{{2, −3, 2, −5}, {1, 5, −4, 1}, {−3, 1, −1/4, 2}, {5, −5, −1, 3}}]
> **b** = {{**−21**}, {**0**}, {**8**}, {**6**}}
> **Ainv**.**b**

It produces the following output:

```
>{{1},{2},{4},{5}}
```

So, the unique solution is

$$\begin{bmatrix} x \\ y \\ z \\ w \end{bmatrix} = \begin{bmatrix} 1 \\ 2 \\ 4 \\ 5 \end{bmatrix}$$

EXAMPLE 7.26 Solve the following linear system of equations:

$$\begin{cases} 3x + 8y + 10z = 15 \\ 7x + y + 2z = 25 \\ -5x + 4y - z = -12 \end{cases}$$

Solution: We can rewrite the system in a matrix form:

$$\begin{bmatrix} 3 & 8 & 10 \\ 7 & 1 & 2 \\ -5 & 4 & -1 \end{bmatrix} \begin{bmatrix} x \\ y \\ z \end{bmatrix} = \begin{bmatrix} 15 \\ 25 \\ -12 \end{bmatrix}$$

First, we find the rank of the coefficient matrix and the augmented matrix using the following code:

> **MatrixRank**[{{3, 8, −10}, {7, 1, 2}, {−5, 2, −1}}]
> **MatrixRank**[{{3, 8, −10, 15}, {7, 1, 2, 25}, {−5, 2, −1, −12}}]

It follows that the rank of the coefficient matrix equals the rank of the augmented matrix:

$$\text{Rank}(\mathbf{A}) = \text{Rank}(\mathbf{A}/\mathbf{b}) = 3$$

Therefore there is a unique solution to the system of linear equations. The code below allows us to find the solution:

```
> Ainv = Inverse[{{3, 8, −10}, {7, 1, 2}, {−5, 2, −1}}]
> b = {{15}, {25}, {−12}}
> Ainv.b
```

It produces the following output:

```
{{3},{2},{1}}
```

Therefore, the system has a unique solution

$$\begin{bmatrix} x \\ y \\ z \end{bmatrix} = \begin{bmatrix} 3 \\ 2 \\ 1 \end{bmatrix}$$

EXAMPLE 7.27 Solve the following linear system of equations:

$$\begin{cases} 2x - y + z = 2 \\ 4x - 2y + 2z = 5 \\ 5x - 4y + 3z = 9 \end{cases}$$

Solution: First, we find the rank of the coefficient matrix and the augmented matrix using the following code:

```
> MatrixRank[{{2, −1, 1}, {4, −2, 2}, {5, −4, 3}}]
> MatrixRank[{{2, −1, 1, 2}, {4, −2, 2, 5}, {5, −4, 3, 9}}]
```

It follows that

$$\text{Rank}(\mathbf{A}) = 2$$
$$\text{Rank}(\mathbf{A}/\mathbf{b}) = 3$$

so, there is no solution to the system of linear equations. The linear system of equations is inconsistent.

7.10 Matlab examples

Matlab can perform just about any matrix operation that Mathematica can. For instance, to define matrix

$$\mathbf{B} = \begin{bmatrix} 3 & -2 & 1 \\ 5 & -1 & 3 \\ 1 & -4 & -2 \end{bmatrix}$$

in Matlab we write

> **B** = [**3** − **2 1**; **5** − **1 3**; **1** − **4** − **2**]

```
       3   -2    1
   B = 5   -1    3
       1   -4   -2
```

To find the inverse and rank of **B**, we write **inv**(**B**) and **Rank**(**B**).

EXAMPLE **7.28** Solve the following linear system of equations:

$$\begin{cases} 2x - y + z = 2 \\ 4x - 2y + 2z = 5 \\ 5x - 4y + 3z = 9 \end{cases}$$

Solution: Find the rank of the coefficient matrix and the augmented matrix using the following code:

> **Rank**([**2** − **1 1**; **4** − **2 2**; **5** − **4 3**])

> **Rank**([**2** − **1 1 2**; **4** − **2 2 5**; **5** − **4 3 9**])

It follows that

$$\text{Rank}(\mathbf{A}) = 2$$

$$\text{Rank}(\mathbf{A}/\mathbf{b}) = 3$$

so, there is no solution to the system of linear equations. The linear system of equations is inconsistent.

8 Optimization of bivariate and multivariate functions

In Chapter 5 we have considered optimization of single-variable functions. Most economic quantities, however, depend not on a single variable, but on several. Such dependences lead to analysis of functions of several variables (multivariate functions). Because of the many variables involved, the usual scalar quantities like derivative and second derivative are replaced, correspondingly, with the gradient vector and the Hessian matrix. After having introduced the necessary technical matrix tools in Chapters 6 and 7, we now have all the necessary prerequisites to deal with the calculus of several variables.

8.1 The Hessian matrix

Let $y = f(x_1, x_2, \ldots, x_n)$ be a scalar-valued C^2 function of n variables defined on an open subset D of $\mathbb{R}^n$. If $\mathbf{x} = (x_1, x_2, \ldots, x_n)$, then

$$y = f(\mathbf{x})$$

and the *vector of partial derivatives* (also known as the *gradient vector*) is given by

$$\frac{\partial f(\mathbf{x})}{\partial \mathbf{x}} = \nabla f = \begin{bmatrix} f_1 \\ f_2 \\ \vdots \\ f_n \end{bmatrix} = \begin{bmatrix} \dfrac{\partial f}{\partial x_1} \\ \dfrac{\partial f}{\partial x_2} \\ \vdots \\ \dfrac{\partial f}{\partial x_n} \end{bmatrix}$$

The dimension of the gradient vector is $n \times 1$. We can think of the gradient vector as the matrix of first derivarives. The *Hessian matrix* (also known as the *Hessian* or *second-derivatives matrix*) of $f(x_1, x_2, \ldots, x_n)$ is computed as

$$\mathbf{H} = \begin{bmatrix} f_{11} & f_{12} & \cdots & f_{1n} \\ f_{21} & f_{22} & \cdots & f_{2n} \\ \vdots & \vdots & \ddots & \vdots \\ f_{n1} & f_{n2} & \cdots & f_{nn} \end{bmatrix}$$

The *Hessian determinant* of $f(x_1, x_2, \ldots, x_n)$ is

$$|\mathbf{H}| = \begin{vmatrix} f_{11} & f_{12} & \cdots & f_{1n} \\ f_{21} & f_{22} & \cdots & f_{2n} \\ \vdots & \vdots & \ddots & \vdots \\ f_{n1} & f_{n2} & \cdots & f_{nn} \end{vmatrix}$$

and it plays a very important role in optimization problems.

The *leading principal minors* of the Hessian matrix are

$$|\mathbf{H}_1| = |f_{11}|$$

$$|\mathbf{H}_2| = \begin{vmatrix} f_{11} & f_{12} \\ f_{21} & f_{22} \end{vmatrix}$$

$$|\mathbf{H}_3| = \begin{vmatrix} f_{11} & f_{12} & f_{13} \\ f_{21} & f_{22} & f_{23} \\ f_{31} & f_{32} & f_{33} \end{vmatrix}$$

and

$$|\mathbf{H}_n| = |\mathbf{H}| = \begin{vmatrix} f_{11} & f_{12} & \cdots & f_{1n} \\ f_{21} & f_{22} & \cdots & f_{2n} \\ \vdots & \vdots & \ddots & \vdots \\ f_{n1} & f_{n2} & \cdots & f_{nn} \end{vmatrix}$$

8.2 Two-variable functions

Consider a function $z = f(x, y) = x^2 + y^2$ defined on an open set D in the xy-plane (Figure 8.1). The function $f(x, y)$ attains its minimum value at an interior point $(x, y) = (0, 0)$ of D. It is seen rather readily that $f(x, y)$ takes only non-negative values and $f(x, y) = 0$ if and only if $x = y = 0$. Thus, we can tell just by looking at the function $f(x, y)$ that it has a unique global minimum at point $(0, 0)$.

8.2.1 Conditions for extrema

Let $z = f(x, y)$ be a C^1 function defined on an open subset D of $\mathbb{R}^2$. A point $p = (x_0, y_0)$ of D is an extremum point of $f(x, y)$ only if the function is stationary at p, that is, the first-order conditions below are satisfied:

$$f_1(x_0, y_0) = f_2(x_0, y_0) = 0$$

These are known as the *first-order necessary conditions* (FOC's) for interior extrema (maxima or minima).

Note that a stationary point of a function of two variables does not have to be an extremum. A stationary point may be a saddle point. Generally, at a saddle point the surface resembles that of a saddle: curves up in one direction, and curves down in another direction (see point $A = (0, 0, 0)$ in Figure 8.2).

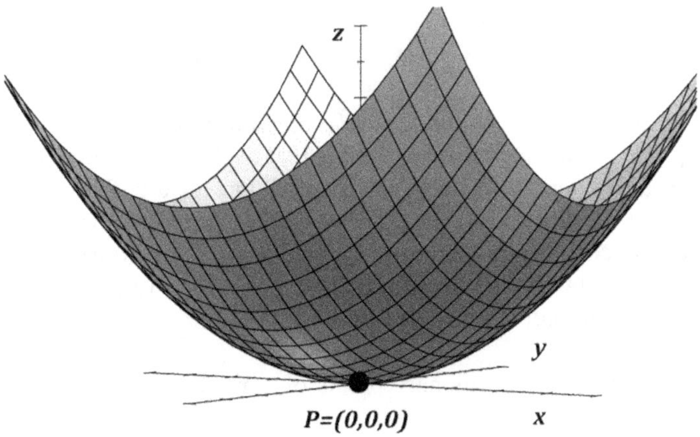

Figure 8.1 $z = x^2 + y^2$.

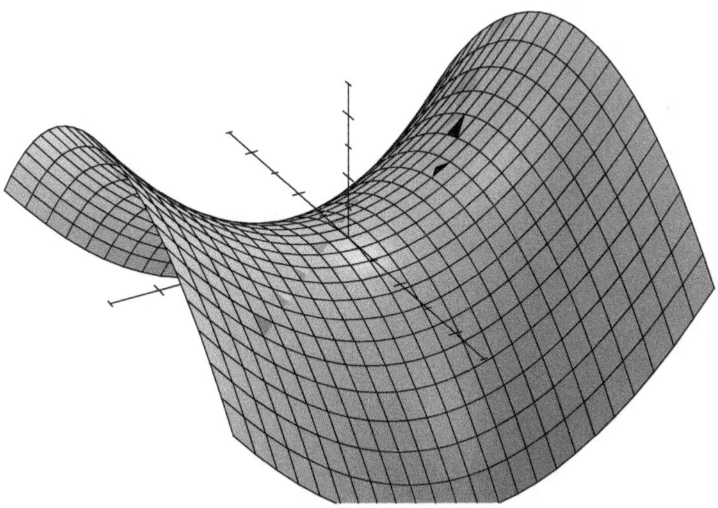

Figure 8.2 $z = x^2 - y^2$.

To sum up, a stationary point $p = (x_0, y_0)$ of a function of two variables can only be

1. a local maximum,
2. a local minimum, or
3. a saddle point.

A simple criterion for checking if a given stationary point $p = (x_0, y_0)$ of a two-variable function $f(x, y)$ defined somewhere on $\mathbb{R}^2$ is a minimum, a maximum or a saddle point is listed below:

(a) $|\mathbf{H}_1(x_0, y_0)|, |\mathbf{H}_2(x_0, y_0)| > 0 \;\Rightarrow p$ is a local minimum;
(b) $|\mathbf{H}_1(x_0, y_0)| < 0, |\mathbf{H}_2(x_0, y_0)| > 0 \Rightarrow p$ is a local maximum;

(c) $|\mathbf{H}_2(x_0, y_0)| < 0 \Rightarrow p$ is a saddle point;
(d) $|\mathbf{H}_2(x_0, y_0)| = 0 \Rightarrow p$ may be a local maximum, a local minimum or a saddle point.

We shall refer to the conditions above as the *second-order conditions* (SOCs).

Often in optimization problems we need to make certain that the local maximum (or minimum) that we have found also happens to be a global maximum (or a global minimum), preferably a unique global maximum. Without knowing whether our objective function is concave or convex, we cannot tell whether the local maximum (or minimum) we have found is also a global maximum (or minimum). We will address this issue in detail in the next section. For now we will focus on finding local extrema only.

EXAMPLE 8.1 Find and classify the stationary points of $z(x, y) = 8x^3 + 2xy - 3x^2 + y^2 + 17$.

Solution: The FOCs for optimization are

$$\begin{cases} \dfrac{\partial z}{\partial x} = 24x^2 + 2y - 6x = 0 \\ \dfrac{\partial z}{\partial y} = 2x + 2y = 0 \end{cases}$$

The system

$$\begin{cases} 24x^2 + 2y - 6x = 0 \\ \qquad\quad 2x + 2y = 0 \end{cases}$$

has exactly two solutions:

$$(x_1^*, y_1^*) = (0, 0)$$

and

$$(x_2^*, y_2^*) = \left(\tfrac{1}{3}, -\tfrac{1}{3}\right)$$

The Hessian determinant of this function is

$$|\mathbf{H}_2| = \begin{vmatrix} 48x - 6 & 2 \\ 2 & 2 \end{vmatrix}$$

The leading principal minors of the Hessian matrix are

$$|\mathbf{H}_1| = |48x - 6|, \quad |\mathbf{H}_2| = \begin{vmatrix} 48x - 6 & 2 \\ 2 & 2 \end{vmatrix}$$

The leading principal minors of the Hessian matrix evaluated at (x_1^*, y_1^*) and (x_2^*, y_2^*) are

$$\begin{aligned} |\mathbf{H}_1(x_1^*, y_1^*)| &= -6 < 0, \quad & |\mathbf{H}_2(x_1^*, y_1^*)| &= -16 < 0 \\ |\mathbf{H}_1(x_2^*, y_2^*)| &= 10 > 0, \quad & |\mathbf{H}_2(x_2^*, y_2^*)| &= 16 > 0 \end{aligned}$$

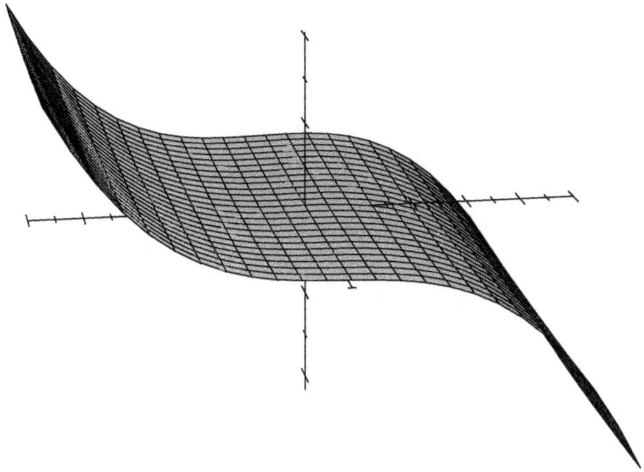

Figure 8.3 $z = 8x^3 + 2xy - 3x^2 + y^2 + 17$.

It follows immediately that $(x_1^*, y_1^*) = (0, 0)$ is a saddle point and $(x_2^*, y_2^*) = \left(\frac{1}{3}, -\frac{1}{3}\right)$ is a local minimum of $z(x, y)$.

The surface at a saddle point may or may not look like a saddle. In this particular example, the saddle point does not give a rise to a saddle-like surface (see Figure 8.3).

EXAMPLE 8.2 A competitive firm produces its output according to the following Cobb–Douglas production function

$$Q = f(K, L) = 400K^{\frac{1}{5}}L^{\frac{1}{5}}$$

The firm pays its labor an hourly wage of $w = 5$ and the price of capital is $r = 15$. The firm sells its output at one dollar per unit. Determine the profit-maximizing level of each input the firm should use.

Solution: The profit function of the firm is:

$$\begin{aligned}\pi(K, L) &= \text{TR} - \text{TC}\\ &= PQ - wL - rK\\ &= 400K^{\frac{1}{5}}L^{\frac{1}{5}} - 5L - 15K\end{aligned}$$

The FOCs for profit maximization are

$$\frac{\partial \pi}{\partial K} = 80K^{-\frac{4}{5}}L^{\frac{1}{5}} - 15 = 0 \qquad (8.1)$$

$$\frac{\partial \pi}{\partial L} = 80K^{\frac{1}{5}}L^{-\frac{4}{5}} - 5 = 0 \qquad (8.2)$$

From equations (8.1) and (8.2) we get

$$\frac{L}{K} = 3$$

or

$$L = 3K$$

Substituting this into equation (8.1) and solving for K yields

$$K^* = \left(\frac{15}{80 \cdot 3^{\frac{1}{5}}}\right)^{-\frac{5}{3}} \approx 23.5$$

$$L^* = 3\left(\frac{15}{8 \cdot 3^{\frac{1}{5}}}\right)^{-\frac{5}{3}} \approx 70.5$$

Now, we check the SOCs for profit maximization. The Hessian for the profit function is

$$|\mathbf{H}| = \begin{vmatrix} \pi_{KK} & \pi_{KL} \\ \pi_{LK} & \pi_{LL} \end{vmatrix}$$

$$= \begin{vmatrix} -64K^{-\frac{9}{5}}L^{\frac{1}{5}} & 16(KL)^{-\frac{4}{5}} \\ 16(KL)^{-\frac{4}{5}} & -64K^{\frac{1}{5}}L^{-\frac{9}{5}} \end{vmatrix}$$

Since

$$|\mathbf{H}_1| = -64K^{-\frac{9}{5}}L^{\frac{1}{5}} < 0$$

and

$$|\mathbf{H}_2| = (KL)^{-\frac{8}{5}}(64^2 - 16^2) > 0$$

the second-order condition for a local maximum is satisfied. The optimal output level is found by substituting K^* and L^* into the production function:

$$Q^* = 400(K^*L^*)^{\frac{1}{5}} \approx 1762 \text{ (units)}$$

Now, let us consider a very similar profit-maximization problem where w, r and P are given as parameters.

EXAMPLE 8.3 A competitive firm produces its output according to the following Cobb–Douglas production function

$$Q = f(K, L) = AK^{\alpha}L^{\beta}, \quad \text{where } A, \alpha, \beta > 0 \text{ and } \alpha = \beta < \frac{1}{2}$$

The firm pays its labor an hourly wage of w and the price of capital is r. The firm sells its output at price P dollars per unit. Determine the profit-maximizing level of each input the firm should use and the optimal level of output.

Solution: There are two endogenous variables (K and L) and six exogenous variables (A, α, β, P, w and r) in this model. Since α equals β, from now on we will use β in place of α.

The profit function of the firm is:

$$\begin{aligned}\pi(K, L) &= \mathrm{TR} - \mathrm{TC} \\ &= PQ - wL - rK \\ &= PAK^{\beta}L^{\beta} - wL - rK\end{aligned}$$

The first-order condition for profit maximization is

$$\frac{\partial \pi}{\partial K} = PA\beta K^{\beta-1}L^{\beta} - r = 0 \tag{8.3}$$

$$\frac{\partial \pi}{\partial L} = PA\beta K^{\beta}L^{\beta-1} - w = 0 \tag{8.4}$$

Dividing (8.3) by (8.4) yields

$$\frac{L}{K} = \frac{r}{w}$$

or

$$L = \frac{r}{w}K$$

Substituting this into (8.3) and solving (8.3) for K yields

$$K^* = \left(\frac{r^{1-\beta}w^{\beta}}{PA\beta}\right)^{\frac{1}{2\beta-1}}$$

Thus, we have

$$L^* = \frac{r}{w}\left(\frac{r^{1-\beta}w^{\beta}}{PA\beta}\right)^{\frac{1}{2\beta-1}}.$$

Let us now check the second-order condition to verify that we have a maximum:

$$\begin{aligned}|\mathbf{H}| &= \begin{vmatrix} \pi_{KK} & \pi_{KL} \\ \pi_{LK} & \pi_{LL} \end{vmatrix} \\ &= \begin{vmatrix} (\beta-1)PA\beta K^{\beta-2}L^{\beta} & PA\beta^2K^{\beta-1}L^{\beta-1} \\ PA\beta^2K^{\beta-1}L^{\beta-1} & (\beta-1)PA\beta K^{\beta}L^{\beta-2} \end{vmatrix}\end{aligned}$$

Since

$$|\mathbf{H}_1| = (\beta - 1)PA\beta K^{\beta-2}L^{\beta} < 0$$

and

$$|\mathbf{H}_2| = (PA\beta)^2(KL)^{2\beta-2}[1 - 2\beta] > 0$$

we conclude that the profit function attains a local maximum at K^* and L^*. Now we can substitute the optimal levels of capital and labor in the production function to find that

$$\begin{aligned} Q^* &= A(K^*)^{\beta}(L^*)^{\beta} \\ &= A\left(\frac{r}{w}\right)^{\beta}\left(\frac{r^{1-\beta}w^{\beta}}{PA\beta}\right)^{\frac{2\beta}{2\beta-1}} \end{aligned}$$

This gives us an expression for Q^* as a function of the exogenous variables A, P, β, w and r.

Example 8.4 A firm has two factories, one in China and one in the US. Each factory produces the same item. Let x and y denote the respective output levels of the two factories. The cost functions for the two factories are

$$C_1(x) = \frac{x^2}{60} - 10x + 250, \quad C_2(y) = \frac{y^3}{150} - 50y - 100$$

The total cost of production is

$$C(x, y) = C_1(x) + C_2(y)$$

Find the values of x and y that minimize the total cost $C(x, y)$.

Solution: Substituting expressions for C_1 and C_2 in the expression for the total cost yields

$$C(x, y) = \frac{x^2}{60} - 10x + \frac{y^3}{150} - 50y + 150$$

The first-order conditions for cost minimization are

$$\begin{aligned} \frac{\partial C}{\partial x} &= \frac{x}{30} - 10 = 0 \\ \frac{\partial C}{\partial y} &= \frac{y^2}{50} - 50 = 0 \end{aligned}$$

The function in question has a unique critical point at $(x^*, y^*) = (300, 50)$. Let us now check the second-order condition to verify that we have a minimum:

$$\begin{aligned} |\mathbf{H}| &= \begin{vmatrix} C_{xx} & C_{xy} \\ C_{yx} & C_{yy} \end{vmatrix} \\ &= \begin{vmatrix} \frac{1}{30} & 0 \\ 0 & \frac{y}{25} \end{vmatrix} \end{aligned}$$

It follows that

$$|\mathbf{H}_1| = \frac{1}{30} > 0$$
$$|\mathbf{H}_2| = \frac{y}{750}$$

In problems like this one we must evaluate the determinant of the Hessian matrix at the critical point. Indeed, while $|\mathbf{H}_1| > 0$ everywhere in the domain of $C(x, y)$, the sign of $|\mathbf{H}_2|$ depends on y. Evaluating the Hessian at the critical point $(x^*, y^*) = (300, 50)$ yields

$$|\mathbf{H}_1(x^*, y^*)| = \frac{1}{30}$$
$$|\mathbf{H}_2(x^*, y^*)| = \begin{vmatrix} \frac{1}{30} & 0 \\ 0 & \frac{50}{25} \end{vmatrix} = \frac{1}{15} > 0$$

Since $|\mathbf{H}_1(x^*, y^*)| > 0$ and $|\mathbf{H}_2(x^*, y^*)| > 0$ it follows that $(x^*, y^*) = (300, 50)$ is a minimum point of the cost function.

8.3 Multivariate functions

8.3.1 Conditions for extrema

Now, let us say we have a function of $n \geq 2$ variables, say $f = f(x_1, x_2, \ldots, x_n)$, defined on an open set D in $\mathbb{R}^n$. Using vector notation, we can write the function in question as $f(\mathbf{x})$, where $\mathbf{x} \in \mathbb{R}^n$. The necessary condition for a point $\mathbf{x}^*$ to be an extremum is

$$f_1(\mathbf{x}^*) = f_2(\mathbf{x}^*) = \cdots = f_n(\mathbf{x}^*) = 0$$

This condition is typically referred to as the *first-order conditions for optimization* (minimization or maximization).

The *second-order sufficient conditions for minimization* are that all leading principal minors of the Hessian matrix are positive:

$$|\mathbf{H}_1(\mathbf{x}^*)|, |\mathbf{H}_2(\mathbf{x}^*)|, \ldots, |\mathbf{H}_n(\mathbf{x}^*)| > 0$$

The *second-order sufficient conditions for maximization* are that all leading principal minors of the Hessian alternate in sign as shown below:

$$|\mathbf{H}_1(\mathbf{x}^*)| < 0; |\mathbf{H}_2(\mathbf{x}^*)| > 0; |\mathbf{H}_3(\mathbf{x}^*)| < 0; \ldots; (-1)^n|\mathbf{H}_n(\mathbf{x}^*)| > 0$$

As in the two-variable case, we need to know the curvature of our objective function to tell whether the local maximum (or minimum) we have found is simultaneously a global maximum (or minimum). We shall routinely use the theorem below to answer that question.

THEOREM 8.1: *Let $f: D \to \mathbb{R}$ be a scalar-valued C^2 function, where D is a convex open subset of $\mathbb{R}^n$. Then:*

(i) f is concave on $D \Leftrightarrow \mathbf{H}(\mathbf{x})$ is negative semidefinite for all $\mathbf{x}$ in D;
(ii) f is strictly concave on $D \Leftrightarrow \mathbf{H}(\mathbf{x})$ is negative definite for all $\mathbf{x}$ in D;
(iii) f is convex on $D \Leftrightarrow \mathbf{H}(\mathbf{x})$ is positive semidefinite for all $\mathbf{x}$ in D;
(iv) f is strictly convex on $D \Leftrightarrow \mathbf{H}(\mathbf{x})$ is positive definite for all $\mathbf{x}$ in D;
(v) f is concave on D and $\nabla f(\mathbf{x}^*) = 0$ for some $\mathbf{x}^*$ in $D \Rightarrow \mathbf{x}^*$ is a global maximum of f;
(vi) f is strictly concave on D and $\nabla f(\mathbf{x}^*) = 0$ for some $\mathbf{x}^*$ in $D \Rightarrow \mathbf{x}^*$ is a unique global maximum of f;
(vii) f is convex on D and $\nabla f(\mathbf{x}^*) = 0$ for some $\mathbf{x}^*$ in $D \Rightarrow \mathbf{x}^*$ is a global minimum of f;
(viii) f is strictly convex on D and $\nabla f(\mathbf{x}^*) = 0$ for some $\mathbf{x}^*$ in $D \Rightarrow \mathbf{x}^*$ is a unique global minimum of f.

EXAMPLE 8.5 Find and classify the stationary points of the following function.

$$z = z(x, y, w) = x^2 + y^2 + w^2$$

Solution: The first-order necessary condition

$$z_x = 2x = 0$$
$$z_y = 2y = 0$$
$$z_w = 2w = 0$$

implies that the only stationary point of the function is $A = (0, 0, 0)$. It is seen rather readily that A is a minimum of the function. However, let us formally show that this is the case. Since

$$|\mathbf{H}| = \begin{vmatrix} z_{xx} & z_{xy} & z_{xw} \\ z_{yx} & z_{yy} & z_{yw} \\ z_{wx} & z_{wy} & z_{ww} \end{vmatrix} = \begin{vmatrix} 2 & 0 & 0 \\ 0 & 2 & 0 \\ 0 & 0 & 2 \end{vmatrix}$$

and

$$|\mathbf{H}_1|, |\mathbf{H}_2|, |\mathbf{H}_3| > 0$$

we conclude that point $A = (0, 0, 0)$ is a local minimum. Since the Hessian is positive definite everywhere, we conclude that point A is a unique global minimum. Recall, that by Theorem 8.1(iv) the underlying function is strictly convex. Hence, by Theorem 8.1(vi) the point A is a unique global minimum.

EXAMPLE 8.6 Find and classify the stationary points of the following function.

$$f(x, y, z) = 5 - 4x + 10y - 2x^2 - 6y^2 + 2xy - z^2$$

Solution: First, we state the FOCs:

$$\nabla f = \begin{bmatrix} -4-4x+2y \\ 10-12y+2x \\ -2z \end{bmatrix} = \mathbf{0}$$

The system of equations above has a unique solution $A = \left(-\frac{7}{11}, \frac{8}{11}, 0\right)$. Thus, the point $A = \left(-\frac{7}{11}, \frac{8}{11}, 0\right)$ is the stationary point of the function. The Hessian of the function is

$$|\mathbf{H}| = \begin{vmatrix} f_{xx} & f_{xy} & f_{xz} \\ f_{yx} & f_{yy} & f_{yz} \\ f_{zx} & f_{zy} & f_{zz} \end{vmatrix} = \begin{vmatrix} -4 & 2 & 0 \\ 2 & -12 & 0 \\ 0 & 0 & -2 \end{vmatrix}$$

Since

$$|\mathbf{H}_1| = -4 < 0$$
$$|\mathbf{H}_2| = 44 > 0$$
$$|\mathbf{H}_3| = -88 < 0$$

we conclude that $f(x, y, z)$ is maximized at point $A = \left(-\frac{7}{11}, \frac{8}{11}, 0\right)$.

EXAMPLE 8.7 Find and classify the stationary points of the following function.

$$f(x, y, z) = -4x + 10y + 2x^2 + 6y^2 - 6xy + 2z^2 + 5$$

Solution: First, we obtain the first-order conditions:

$$\nabla f = \begin{bmatrix} -4+4x-6y \\ 10+12y-6x \\ 4z \end{bmatrix} = \mathbf{0}$$

The system of equations above has a unique solution $A = \left(-1, -\frac{4}{3}, 0\right)$. Note that point $A = \left(-1, -\frac{4}{3}, 0\right)$ is the only stationary point of the function. The Hessian matrix of the function is

$$|\mathbf{H}| = \begin{vmatrix} f_{xx} & f_{xy} & f_{xz} \\ f_{yx} & f_{yy} & f_{yz} \\ f_{zx} & f_{zy} & f_{zz} \end{vmatrix} = \begin{vmatrix} 4 & -6 & 0 \\ -6 & 12 & 0 \\ 0 & 0 & 4 \end{vmatrix}$$

Since

$$|\mathbf{H}_1| = 4 > 0$$
$$|\mathbf{H}_2| = 12 > 0$$
$$|\mathbf{H}_3| = 48 > 0$$

we conclude that $f(x, y, z)$ is minimized at point $A = \left(-1, -\frac{4}{3}, 0\right)$.

EXAMPLE 8.8 Find and classify the stationary points of the following function.

$$f = f(x, y, z, w) = y + 12z + 3w - x^2 - y^2 - z^2 - 0.5w^2 + xy + 17$$

Solution: First, we obtain the first-order conditions:

$$\nabla f = \begin{bmatrix} -2x + y \\ 1 - 2y + x \\ 12 - 2z \\ 3 - w \end{bmatrix} = \mathbf{0}_{4\times 1}$$

The system of equations above has one solution

$$A = (x = \tfrac{1}{3},\ y = \tfrac{2}{3},\ z = 6,\ w = 3)$$

The Hessian matrix of the function is given by

$$|\mathbf{H}| = \begin{vmatrix} f_{xx} & f_{xy} & f_{xz} & f_{xw} \\ f_{yx} & f_{yy} & f_{yz} & f_{yw} \\ f_{zx} & f_{zy} & f_{zz} & f_{zw} \\ f_{wx} & f_{wy} & f_{wz} & f_{ww} \end{vmatrix} = \begin{vmatrix} -2 & 1 & 0 & 0 \\ 1 & -2 & 0 & 0 \\ 0 & 0 & -2 & 0 \\ 0 & 0 & 0 & -1 \end{vmatrix}$$

Since

$$|\mathbf{H}_1| = -2$$
$$|\mathbf{H}_2| = 3$$
$$|\mathbf{H}_3| = -6$$
$$|\mathbf{H}_4| = 6$$

we conclude that $f(x, y, z, w)$ is maximized at point $A = \left(\frac{1}{3}, \frac{2}{3}, 6, 3\right)$. Since the Hessian is negative definite everywhere, we conclude that point A is a unique global maximum.

8.3.2 A three-factory firm

Consider a firm that operates three separate factories with total cost functions given by

$$C_i = C_i(q_i), \quad \text{where } i = 1, 2, 3$$

and q_i is the output level at factory i and the marginal cost curves are upward-sloping ($C_i'' > 0$ for $i = 1, 2, 3$).

Let us assume that the firm sells all of its output to a single competitive market at a price P and has a total revenue function that depends on the total level of output Q:

$$R = R(Q), \quad \text{where } Q = q_1 + q_2 + q_3$$

Thus, the firm's profit is

$$\pi = \text{TR} - \text{TC} = R(Q) - [C_1(q_1) + C_2(q_2) + C_3(q_3)]$$

The FOCs for profit maximization are given by

$$\frac{\partial \pi}{\partial q_1} = R'(Q) - C_1'(q_1) = 0$$
$$\frac{\partial \pi}{\partial q_2} = R'(Q) - C_2'(q_2) = 0$$
$$\frac{\partial \pi}{\partial q_3} = R'(Q) - C_3'(q_3) = 0$$

It follows readily from the FOCs that

$$C_1'(q_1^*) = C_2'(q_2^*) = C_3'(q_3^*) = R'(Q^*)$$

In other words, to maximize profits the factories must operate at the same level of marginal cost. Further, the level of marginal cost at which the three factories operate must equal the marginal revenue from their combined output.

The Hessian matrix for the profit function is

$$\mathbf{H} = \begin{bmatrix} \pi_{11} & \pi_{12} & \pi_{13} \\ \pi_{21} & \pi_{22} & \pi_{23} \\ \pi_{31} & \pi_{32} & \pi_{33} \end{bmatrix} = \begin{bmatrix} R'' - C_1'' & R'' & R'' \\ R'' & R'' - C_2'' & R'' \\ R'' & R'' & R'' - C_3'' \end{bmatrix}$$

Recall that in perfect competition the price is fixed ($P = MR = R'$), which means that

$$R'' = 0$$

Thus, the Hessian matrix for the profit function can be written as

$$\mathbf{H} = \begin{bmatrix} -C_1'' & 0 & 0 \\ 0 & -C_2'' & 0 \\ 0 & 0 & -C_3'' \end{bmatrix}$$

The principal minors of the Hessian matrix are alternating in sign

$$|\mathbf{H}_1| = -C_1'' < 0$$
$$|\mathbf{H}_2| = C_1''C_2'' > 0$$
$$|\mathbf{H}_3| = -C_1''C_2''C_3'' < 0$$

and imply that the profit function is maximized at (q_1^*, q_2^*, q_3^*).

8.3.3 Bordered Hessian

A bordered Hessian is employed for the second-derivative test in certain constrained optimization problems. Suppose we wish to maximize the function

$$f = f(x_1, x_2, \ldots, x_n)$$

subject to a constraint

$$g = g(x_1, x_2, \ldots, x_n)$$

Then the bordered Hessian is defined as

$$\overline{\mathbf{H}} = \begin{bmatrix} 0 & g_1 & g_2 & \cdots & g_n \\ g_1 & f_{11} & f_{12} & \cdots & f_{1n} \\ g_2 & f_{21} & f_{22} & \cdots & f_{2n} \\ \vdots & \vdots & \vdots & \ddots & \vdots \\ g_n & f_{n1} & f_{n2} & \cdots & f_{nn} \end{bmatrix}$$

8.4 Optimization with one constraint

To optimize a function of n variables $f(x_1, x_2, \ldots, x_n)$ subject to the constraint $g(x_1, x_2, \ldots, x_n) = 0$, we set up a *Lagrangian function* $\mathcal{L}$ as follows:

$$\mathcal{L} = f(x_1, x_2, \ldots, x_n) - \lambda g(x_1, x_2, \ldots, x_n)$$

where λ is a scalar known as the Lagrangian multiplier. Then we optimize the function $\mathcal{L}$ using standard techniques of calculus. We solve the system of equations below to find the critical values of $\mathcal{L}$:

$$\begin{cases} \dfrac{\partial \mathcal{L}}{\partial x_1} = 0 \\ \dfrac{\partial \mathcal{L}}{\partial x_2} = 0 \\ \quad\vdots \\ \dfrac{\partial \mathcal{L}}{\partial x_n} = 0 \\ \dfrac{\partial \mathcal{L}}{\partial \lambda} = 0 \end{cases}$$

Notice that the last equation in the system in fact gives us the original constraint:

$$\frac{\partial \mathcal{L}}{\partial \lambda} = -g(x_1, x_2, \ldots, x_n) = 0$$

To check the second-order conditions we form a bordered Hessian matrix

$$\overline{\mathbf{H}} = \begin{bmatrix} 0 & g_1 & g_2 & \cdots & g_n \\ g_1 & \mathcal{L}_{11} & \mathcal{L}_{12} & \cdots & \mathcal{L}_{1n} \\ g_2 & \mathcal{L}_{21} & \mathcal{L}_{22} & \cdots & \mathcal{L}_{2n} \\ \vdots & \vdots & \vdots & \ddots & \vdots \\ g_n & \mathcal{L}_{n1} & \mathcal{L}_{n2} & \cdots & \mathcal{L}_{nn} \end{bmatrix}$$

and employ the following leading principal minors of $\overline{\mathbf{H}}$:

$$|\overline{\mathbf{H}}_3|, |\overline{\mathbf{H}}_4|, |\overline{\mathbf{H}}_5| \text{ and so on}$$

The critical point $\mathbf{x}^* = (x_1^*, x_2^*, \ldots, x_n^*)$ is a *constrained local maximum*, if the signs of the minors are alternating as follows: $|\overline{\mathbf{H}}_3(\mathbf{x}^*)| > 0$, $|\overline{\mathbf{H}}_4(\mathbf{x}^*)| < 0$, $|\overline{\mathbf{H}}_5(\mathbf{x}^*)| > 0$, and so forth.

The point $x^* = (x_1^*, x_2^*, \ldots, x_n^*)$ is a *constrained local minimum*, if the signs of the minors are negative: $|\overline{\mathbf{H}}_3(\mathbf{x}^*)| < 0$, $|\overline{\mathbf{H}}_4(\mathbf{x}^*)| < 0$, $|\overline{\mathbf{H}}_5(\mathbf{x}^*)| < 0$, and so on.

If neither one of the conditions above hold, the function must be examined numerically (or otherwise) at the critical point.

Example 8.9 Find the minimum of function $f(x, y) = x + y$ subject to the constraint $x^2 + y^2 = 1$.

Solution: Set up the Lagrangian function

$$\mathcal{L} = x + y - \lambda(x^2 + y^2 - 1)$$

Solve the simultaneous equations for x, y and λ:

$$\frac{\partial \mathcal{L}}{\partial x} = 1 - 2\lambda x = 0$$
$$\frac{\partial \mathcal{L}}{\partial y} = 1 - 2\lambda y = 0$$
$$\frac{\partial \mathcal{L}}{\partial \lambda} = -x^2 - y^2 + 1 = 0$$

From the system of equations above we obtain two optimal points

$$\mathbf{z}_1 = (x_1^*, y_1^*, \lambda_1^*) = \left(\frac{\sqrt{2}}{2}, \frac{\sqrt{2}}{2}, \frac{\sqrt{2}}{2}\right)$$
$$\mathbf{z}_2 = (x_2^*, y_2^*, \lambda_2^*) = \left(-\frac{\sqrt{2}}{2}, -\frac{\sqrt{2}}{2}, -\frac{\sqrt{2}}{2}\right)$$

The bordered Hessian looks like

$$\overline{\mathbf{H}} = \begin{bmatrix} 0 & -2x & -2y \\ -2x & -2\lambda & 0 \\ -2y & 0 & -2\lambda \end{bmatrix}$$

The only bordered Hessian minor we care about is $|\overline{\mathbf{H}}_3|$, which is

$$|\overline{\mathbf{H}}_3| = 8\lambda(x^2 + y^2)$$

Evaluating $|\overline{\mathbf{H}}_3|$ at the critical points $\mathbf{z}_1$ and $\mathbf{z}_2$ we get

$$|\overline{\mathbf{H}}_3(\mathbf{z}_1)| = 8\left(\frac{\sqrt{2}}{2}\right)\left(\left(\frac{\sqrt{2}}{2}\right)^2 + \left(\frac{\sqrt{2}}{2}\right)^2\right) > 0$$

$$|\overline{\mathbf{H}}_3(\mathbf{z}_2)| = 8\left(-\frac{\sqrt{2}}{2}\right)\left(\left(-\frac{\sqrt{2}}{2}\right)^2 + \left(-\frac{\sqrt{2}}{2}\right)^2\right) < 0$$

Thus, $\mathbf{z}_1$ is a constrained maximum and $\mathbf{z}_2$ is a constrained minimum of $f(x, y) = x + y$.

Without the Lagrange method, we would have had to solve the constraint for one of the variables, say y, leading to a square root function. This would have made the solution process much more complicated.

EXAMPLE 8.10 Consider a Cobb–Douglas production function

$$q(L, K) = 30K^{0.3}L^{0.7}$$

where q is the production level, K the quantity of capital, and L the amount of labor. Suppose that the firm is interested in increasing its market share by maximizing production. The budget constraint of the firm is

$$5K + 4L = 100$$

Find the optimal levels of capital and labor that maximize production.

Solution: The Lagrangian function in this example is

$$\mathcal{L}(L, K, \lambda) = 30K^{0.3}L^{0.7} - \lambda(5K + 4L - 100)$$

and

$$g(L, K) = 5K + 4L - 100$$

Setting up the first-order conditions for the Lagrangian function, we find

$$\begin{cases} \dfrac{\partial \mathcal{L}}{\partial L} = 21K^{0.3}L^{-0.3} - 4\lambda = 0 \\ \dfrac{\partial \mathcal{L}}{\partial K} = 9K^{-0.7}L^{0.7} - 5\lambda = 0 \\ \dfrac{\partial \mathcal{L}}{\partial \lambda} = -5K - 4L + 100 = 0 \end{cases}$$

From the first two equations we can find the ratio of labor to capital:

$$\frac{L}{K} = \frac{21 \cdot 5}{9 \cdot 4} = \frac{35}{12}$$

and

$$L = \frac{35}{12}K$$

It follows that

$$(L^*, K^*, \lambda^*) = (17.5, 6, 3.8)$$

and the bordered Hessian is

$$\overline{\mathbf{H}} = \begin{bmatrix} 0 & 4 & 5 \\ 4 & -6.3K^{0.3}L^{-1.3} & 6.3K^{-0.7}L^{-0.3} \\ 5 & 6.3K^{-0.7}L^{-0.3} & -6.3K^{-1.7}L^{0.7} \end{bmatrix}$$

with

$$|\overline{\mathbf{H}}_3| = 6.3(40K^{-0.7}L^{-0.3} + 16K^{-1.7}L^{0.7} + 2540K^{0.3}L^{-1.3}) > 0$$

Therefore the critical point

$$(L^*, K^*) = (17.5, 6)$$

is a constrained local maximum of the function.

PROBLEMS FOR SECTION 8.4

1. A perfectly competitive firm produces two types of goods, A and B. The monthly cost of producing a units of good A and b units of good B is

$$C(a, b) = a^2 + b^2 + 2a + 3b + 10$$

The firm sells its output at \$102 per unit of A and \$53 per unit of B. Find the profit-maximizing production levels and compute the firm's maximized profit.

2. Suppose that a perfectly competitive firm operates two plants with the following total cost functions

$$C_1 = 2aq_1^2 + 3bq_1 \quad \text{and} \quad C_2 = 2cq_2^2 + 3dq_2$$

and sells its output at a fixed price P.

(a) Find the optimal output level and firm's maximized profit.

(b) Find $\frac{\partial q_1^*}{\partial P}$ and $\frac{\partial q_2^*}{\partial P}$.

3. A perfectly competitive firm produces three types of goods, A, B and D. The weekly cost of producing a units of good A, b units of good B and d units of good D is

$$C(a, b, d) = a^2 + b^2 + d^2 + 2a + 3b + 4d + 10$$

The firm sells its output at \$102 per unit of A, \$53 per unit of B and \$44 per unit of D. Find the profit-maximizing weekly production levels and compute the maximized profit.

4. A perfectly competitive firm operates three plants with the following total cost functions

$$C_1 = 3aq_1^2 - 4bq_1 + 15$$
$$C_2 = 5cq_2^2 - 6q_2 + b$$
$$C_2 = abcq_3^3 + q_3 + abc$$

and sells its output at a fixed price P.

(a) Find the optimal output level and firm's maximized profit.

(b) Find $\frac{\partial q_1^*}{\partial P}$, $\frac{\partial q_2^*}{\partial P}$ and $\frac{\partial q_3^*}{\partial P}$.

5. Consider a monopolist firm that faces a demand function

$$P = 20 - Q$$

and operates two factories with the given total cost functions

$$C_1 = 2aq_1^2 + b \quad \text{and} \quad C_2 = 2bq_2^2 + c$$

(a) Find the optimal output level.

(b) Find $\frac{\partial q_1^*}{\partial b}$ and $\frac{\partial q_2^*}{\partial a}$.

8.5 Matlab example

Just like Mathematica, Matlab is capable of producing high-quality two- and three-dimensional graphs. The following code in Matlab

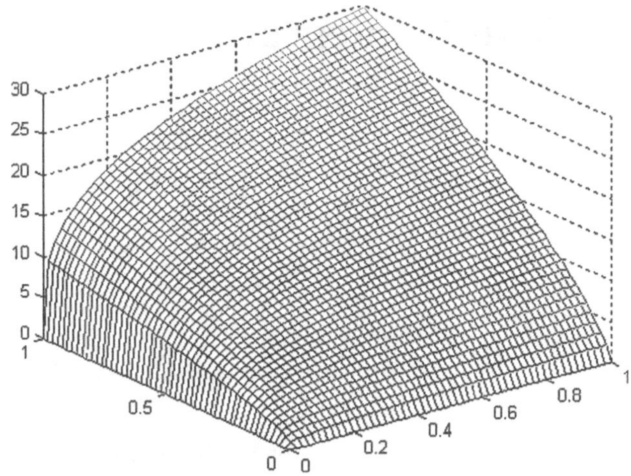

Figure 8.4

>[**x**, **y**] = **meshgrid**(**0**: **0.02**: **1**);

>**z** = **30*****x**.^**0.3**.***y**.^**0.7**;

>**meshc**(**x**, **y**, **z**)

produces a three-dimensional graph of the Cobb–Douglas production function (see Figure 8.4) employed in Example 8.10.

9 Indefinite and definite integrals

Integrals play a twofold role in calculus. The so-called indefinite integral is an operation inverse to differentiation: the integral of function $f(x)$ is a function whose derivative gives us $f(x)$. We have seen the importance of derivatives throughout the textbook, and naturally the inverse operation also plays an important role in mathematical and economic analysis. Later, in Chapter 12, we study relations between quantities and their rates of change – the so-called differential equations. Solving differential equations is impossible without indefinite integrals.

The other type of integral is the definite integral. Geometrically a definite integral represents the area under a curve. But in applications the meaning of integration is totaling continuous quantities. For instance, to reconstruct profit from its rate of change we would integrate the latter function. Another application of integrals is in studying consumer–producer surplus.

9.1 Indefinite integrals

If $f(x)$ and $F(x)$ are some functions of x such that

$$F'(x) = f(x)$$

then $F(x)$ is called an antiderivative of $f(x)$.

For example, the function $F_1(x) = 3x^2 + 5$ is an antiderivative of $f(x) = 6x$. Note that $F_1(x) = 3x^2 + 5$ is not the only antiderivative of $f(x)$. In fact, the function $f(x) = 6x$ has infinitely many antiderivatives and all antiderivatives of $f(x)$ are functions of the form

$$F(x) = 3x^2 + C$$

where C is a constant. We will refer to $F(x) = 3x^2 + C$ as the general antiderivative of $f(x)$ and we will write

$$\int 6x\,dx = 3x^2 + C$$

to indicate that. The symbol $\int$ is referred to as the *integral sign*, $6x$ is the *integrand* and C is the *constant of integration.* The expression $\int 6x\,dx$ is known as the indefinite integral of $f(x) = 6x$.

THEOREM 9.1: *Let $f(x)$ be a differentiable function of x, and k, n and C be some constants. Then*

(Ind1) $\displaystyle\int x^n\,dx = \frac{x^{n+1}}{n+1} + C \quad (n \neq -1)$

(Ind2) $\displaystyle\int x^{-1}\,dx = \ln|x| + C$

(Ind3) $\displaystyle\int dx = x + C$

(Ind4) $\displaystyle\int e^{ku}\,du = \frac{1}{k}e^{ku} + C \quad (k \neq 0)$

(Ind5) $\displaystyle\int k^x\,dx = \frac{k^x}{\ln k} + C \quad (k > 0 \text{ and } k \neq 1)$

(Ind6) $\displaystyle\int kf(x)\,dx = k\int f(x)\,dx$

(Ind7) $\displaystyle\int [f(x) \pm g(x)]\,dx = \int f(x)\,dx \pm \int g(x)\,dx$

(Ind8) $\displaystyle\int -f(x)\,dx = -\int f(x)\,dx$

Observe, that **Ind1** (with $n = 0$) implies **Ind3**, and **Ind4** (with $k = 1$) implies

$$\int e^u\,du = e^u + C$$

EXAMPLE 9.1 Evaluate the following integrals

(a) $\displaystyle\int y^{0.5}\,dy$

(b) $\displaystyle\int (w^{0.3} + w^{0.5})\,dw$

(c) $\displaystyle\int (5t^2 + 3t^5)\,dt$

Solution:

(a) $\displaystyle\int y^{0.5}\,dy = \frac{y^{1.5}}{1.5} + C$

(b)
$$\int (w^{0.3}+w^{0.5})\,dw = \int w^{0.3}\,dw + \int w^{0.5}\,dw$$
$$= \frac{w^{1.3}}{1.3} + \frac{w^{1.5}}{1.5} + C$$

(c)
$$\int (5x^2+3x^4)\,dx = 5\int x^2\,dx + 3\int x^4\,dx$$
$$= \frac{5x^3}{3} + \frac{3x^5}{5} + C_1 + C_2$$
$$= \frac{5x^3}{3} + \frac{3x^5}{5} + C.$$

In Example 9.1(c) note that, since C_1 and C_2 are constants, their sum $C_1 + C_2$ is also a constant.

Sometimes it may be useful to verify the answer by differentiating the general antiderivative function.

EXAMPLE 9.2 Evaluate the following integrals.

(a) $\int (y^{1.5} + y)\,dy$

(b) $\int (5w^2 + 7w^3)\,dw$

(c) $\int (5t^5 + 3t^3 + 17)\,dt$

Solution: To evaluate the integrals above we shall apply **Ind1**, **Ind6** and **Ind7**.

(a) $\int (y^{1.5} + y)\,dy = \frac{y^{2.5}}{2.5} + \frac{y^2}{2} + C$

(b) $\int (5w^2 + 7w^3)\,dw = \frac{5w^3}{3} + \frac{7w^4}{4} + C$

(c) $\int (5t^5 + 3t^3 + 17)\,dt = \frac{5t^6}{6} + \frac{3t^4}{4} + 17t + C$

EXAMPLE 9.3 Evaluate the following integrals.

(a) $\int \left(\frac{2}{y}\right) dy$

(b) $\int 2w^5(w^2 + w^3)\,dw$

(c) $\int e^{2t}\,dt$

Solution:

(a) $\int \left(\frac{2}{y}\right) dy = 2\ln|y| + C$

(b) $\int 2w^5(w^2 + w^3)\,dw = \frac{2w^8}{8} + \frac{2w^9}{9} + C$

(c) $\int e^{2t}\,dt = \frac{e^{2t}}{2} + C$

PROBLEMS FOR SECTION 9.1

1. $\int 3\,dx$

2. $\int 3^{-1}\,dx$

3. $\int 2x^3\,dx$

4. $\int \frac{4}{3}x^3\,dx$

5. $\int \frac{1}{x^4}\,dx$

6. $\int \frac{1}{x^{\frac{4}{7}}}\,dx$

7. $\int 2x^{-5}\,dx$

8. $\int (t^2 + 2t + 7)\,dt$

9. $\int (e^a + e^x)\,dx$

10. $\int \left(\frac{3x^3}{4} + \frac{2x^2}{3} + 7\right) dx$

11. $\int \sqrt{\sqrt{x\sqrt{x\sqrt{x\sqrt{x}}}}}\,dx$

12. $\int \frac{(y-3)^2}{\sqrt{y}}\,dy$

13. $\int \left(\frac{2}{\sqrt{x}} + 3\right)^2 dx$

14. $\int (e^x + e^{2x})\, dx$

15. $\int e^x (e^{2x} + e^{3x})\, dx$

16. $\int (2w + 3)^2\, dw$

17. $\int 0\, dx$

18. $\int 23\, dz$

19. $\int (a + b + x)\, dx$

20. $\int \frac{(x^3 + 3)^2}{x^3}\, dx$

21. $\int \left(\frac{2y^4}{5} + \frac{y^2}{3}\right) dy$

9.2 Integration by substitution and integration by parts

There are circumstances when an indefinite integral can be computed using **Ind1** through **Ind8** of the previous section, but the process is very tedious. Consider, for example, the following indefinite integral:

$$\int 6x(3x^2 + 3)^5\, dx$$

The integrand in this example is not easily multiplied out. In cases like this we shall use *integration by substitution* or the *substitution rule*:

$$\textbf{(Int1)} \int f(u) \frac{du}{dx}\, dx = \int f(u)\, du = F(u) + c$$

This rule is the integral-calculus counterpart of the chain rule.

EXAMPLE 9.4 Compute the following integral:

$$\int 6x(3x^2 + 3)^5\, dx$$

Solution: Here, we shall utilize the substitution rule. Let $u = 3x^2 + 3$; then $\frac{du}{dx} = 6x$. Hence, the integral can be rewritten as

$$\int 6x(3x^2+3)^5\,dx = \int u^5 \frac{du}{dx}\,dx = \int u^5\,du$$
$$= \frac{u^6}{6} + C = \frac{1}{6}(3x^2+3)^6 + C$$

Example 9.5 Compute the following integral:

$$\int 180x(3x^2+3)^5\,dx$$

Solution: Again, we will utilize the substitution rule. Let $u = 3x^2 + 3$; then $\frac{du}{dx} = 6x$. Hence,

$$\int 180x(3x^2+3)^5\,dx = \int 30u^5 \frac{du}{dx}\,dx = 30\int u^5\,du$$
$$= 30\left(\frac{u^6}{6} + C_1\right) = 5u^6 + 30C_1$$
$$= 5(3x^2+3)^6 + C$$

Example 9.6 Compute the following integral:

$$\int 2e^{x+3}\,dx$$

Solution: We shall use the substitution rule here. Let $u = x + 3$; then $\frac{du}{dx} = 1$ and

$$\int 2e^{x+3}\,dx = \int 2e^u\,dx = 2\int e^u\,du$$
$$= 2\left(e^u + C_1\right) = 2e^{x+3} + C$$

Integration by substitution is a very important general method for integrating certain combinations of functions. Another commonly used rule for integrating complex (i.e. difficult) functions is called integration by parts. Let u and v be differentiable functions of x. Then the rule for *integration by parts* is

$$\textbf{(Int2)} \quad \int uv'\,dx = uv - \int u'v\,dx$$

Example 9.7 Compute the following integral:

$$\int 2x(x-2)^2\,dx$$

Solution: Let $u=2x$ and $v=\frac{1}{3}(x-2)^3$, implying that $u'=2$ and $v'=(x-2)^2$. Hence,

$$\begin{aligned}\int 2x(x-2)^2\,dx &= \int uv'\,dx\\ &= uv-\int u'v\,dx\\ &= \frac{2x}{3}(x-2)^3-\int \frac{2}{3}(x-2)^3\,dx\\ &= 2\left(\frac{x}{3}(x-2)^3-\frac{1}{12}(x-2)^4+C_1\right)\\ &= \frac{2x}{3}(x-2)^3-\frac{1}{6}(x-2)^4+C\end{aligned}$$

EXAMPLE 9.8 Compute the following integral:

$$\int x(x+3)^5\,dx$$

Solution: Let $u=x$ and $v=\frac{1}{6}(x+3)^6$, implying that $u'=1$ and $v'=(x+3)^5$. Hence,

$$\begin{aligned}\int x(x+3)^5\,dx &= \frac{x}{6}(x+3)^6-\int \frac{1}{6}(x+3)^6\,dx\\ &= \frac{x}{6}(x+3)^6-\frac{1}{42}(x+3)^7+C\end{aligned}$$

PROBLEMS FOR SECTION 9.2

1. $\int 4x(3x^2+3)^5\,dx$
2. $\int x^2(3x^3+3)^5\,dx$
3. $\int x(3x^2+3)^3\,dx$
4. $\int (3x+2)^5\,dx$
5. $\int (17x+2)^{15}\,dx$
6. $\int (23x+2)^8\,dx$

7. $\int 2x^{-5}\,dx$

8. $\int 2t(t^2+7)^7\,dt$

9. $\int xe^x\,dx$

10. $\int 8\sqrt{2x+3}\,dx$

11. $\int \sqrt{x+3}\,dx$

12. $\int 32\sqrt{4x+12}\,dx$

13. $\int x^2e^x\,dx$

14. $\int xe^{2x}\,dx$

15. $\int x^2e^{3x}\,dx$

16. $\int \frac{\ln x}{x}\,dx$

17. $\int 5xe^{7x}\,dx$

18. $\int \frac{x^3e^{5x}}{3}\,dz$

19. $\int x\ln x\,dx$

20. $\int w^3e^A\,dw$

21. $\int (15t+1)\,dy$

9.3 Definite integrals

9.3.1 An area problem

In Figure 9.1 we display a region A that is bounded above by the graph of a continuous real-valued function $f(x)$ and bounded below by the x-axis. How do we compute the area of A? To accomplish this, let us split the interval $[a,b]$ into $n=4$ non-overlapping equal intervals:

$$[x_0,x_1],\quad [x_1,x_2],\quad [x_2,x_3]\quad \text{and}\quad [x_3,x_4]$$

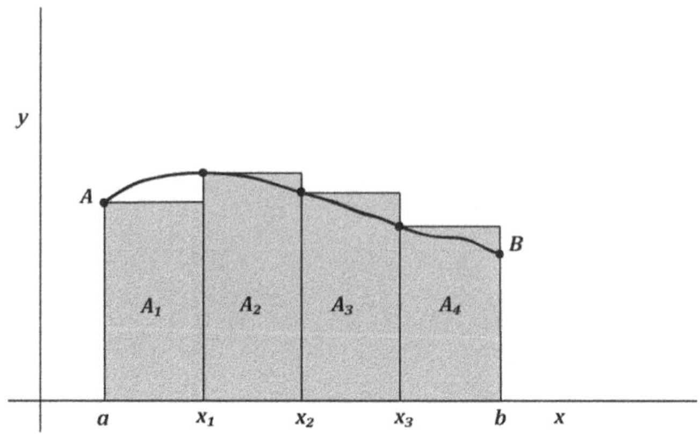

Figure 9.1

where

$$a = x_0 < x_1 < x_2 < x_3 < x_4 = b$$

and the length of each interval is Δx. Now let

$$A_1 = \Delta x \cdot f(x_0)$$
$$A_2 = \Delta x \cdot f(x_1)$$
$$A_3 = \Delta x \cdot f(x_2)$$
$$A_4 = \Delta x \cdot f(x_3)$$

Then, we have

$$A_1 + A_2 + A_3 + A_4 = \sum_{i=0}^{3} f(x_i) \cdot \Delta x = \sum_{i=0}^{n-1} f(x_i) \cdot \Delta x$$

The sum above approximates, albeit poorly, the area of region A. To get a better approximation of the area A, we should divide the interval $[a, b]$ into a much larger number of sub-intervals. The following limit (if it exists)

$$\lim_{n\to\infty} \sum_{i=0}^{n-1} f(x_i) \cdot \Delta x$$

would be equal to the area of A and is written as

$$\int_a^b f(x)\,dx$$

that is

$$\lim_{n\to\infty}\sum_{i=0}^{n-1} f(x_i)\cdot\Delta x = \int_a^b f(x)\,dx$$

The expression

$$\int_a^b f(x)\,dx$$

is referred to as the "integral of $f(x)$ from a to b." The integral

$$\int_a^b f(x)\,dx$$

is also known as the Riemann integral.

Recall that we have assumed in the beginning that $f(x)$ is a continuous real-valued function defined on interval $[a, b]$. Now, in addition to that, let us assume that there is a continuous function $F(x)$ defined on $[a, b]$ with $F'(x) = f(x)$ for all x in (a, b). Then

$$\int_a^b f(x)\,dx = F(b) - F(a)$$

The difference $F(b) - F(a)$ is usually denoted by $F(x)|_a^b$. Thus,

$$\int_a^b f(x)\,dx = F(x)\Big|_a^b$$

Several properties of the definite integral can be derived directly from the definition of the integral. Let $f(x)$, $f_1(x)$ and $f_2(x)$ be continuous functions defined on an interval $[a, b]$ that contains point c and let k_1 and k_2 be some constants. Then we have the following:

(Def1) $\displaystyle\int_a^a f(x)\,dx = 0$

(Def2) $\displaystyle\int_a^b f(x)\,dx = -\int_b^a f(x)\,dx$

(Def3) $\displaystyle\int_a^b kf(x)\,dx = k\int_a^b f(x)\,dx$

(Def4) $\displaystyle\int_a^b [k_1 f_1(x) + k_2 f_2(x)]\,dx = k_1\int_a^b f_1(x)\,dx + k_2\int_a^b f_2(x)\,dx$

(Def5) $\displaystyle\int_a^b f(x)\,dx = \int_a^c f(x)\,dx + \int_c^b f(x)\,dx$

Note that **Def5** can be extended to any number of functions.

EXAMPLE 9.9 Evaluate

$$\int_0^1 x^2\,dx$$

Solution:

$$\int_0^1 x^2\,dx = \left(\frac{x^3}{3}\right)\Bigg|_0^1 = \frac{1^3}{3} - \frac{0^3}{3} = \frac{1}{3}$$

EXAMPLE 9.10 To evaluate

$$\int_0^5 \sqrt{x+5}\,dx$$

we need a function with derivative

$$\sqrt{x+5}$$

It is seen rather readily that

$$\frac{d}{dx}[(x+5)^{\frac{3}{2}}] = \frac{3}{2}(x+5)^{\frac{1}{2}} = \frac{3}{2}\sqrt{x+5}$$

and therefore

$$\frac{d}{dx}\left[\frac{2}{3}(x+5)^{\frac{3}{2}}\right] = \sqrt{x+5}$$

This means that

$$\begin{aligned}\int_0^5 \sqrt{x+5}\,dx &= \left[\frac{2}{3}(x+5)^{\frac{3}{2}}\right]\Bigg|_0^5 \\ &= \frac{2}{3}(10)^{\frac{3}{2}} - \frac{2}{3}(5)^{\frac{3}{2}}\end{aligned}$$

9.3.2 Consumer and producer surplus

EXAMPLE 9.11 Suppose that the demand curve is

$$P = f(Q) = 40 - 0.05Q$$

and the supply curve is

$$P = g(Q) = 0.1Q + 10$$

Find the equilibrium price and compute the consumer and producer surplus.

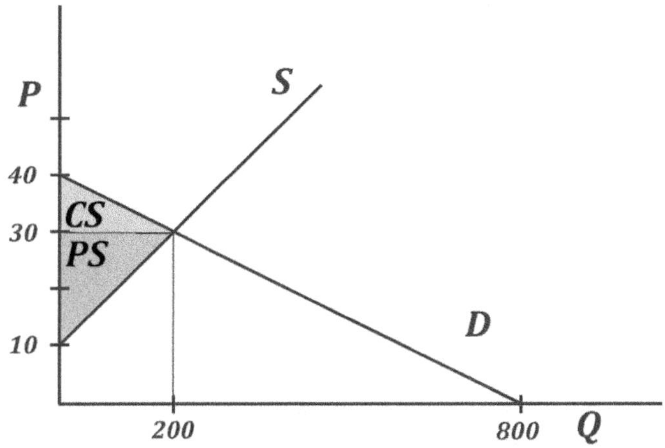

Figure 9.2

Solution: The equilibrium quantity is determined by the equation

$$40 - 0.05Q^* = 0.1Q^* + 10$$

which gives $Q^* = 200$. Then

$$P^* = 0.1Q^* + 10 = 30$$

Figure 9.2 illustrates the consumer surplus (CS) and producer surplus (PS) and shows the equilibrium point. It follows readily that

$$\text{CS} = \int_0^{200} [40 - 0.05Q - 30]\,dQ = \int_0^{200} [10 - 0.05Q]\,dQ = 1000$$

and

$$\text{PS} = \int_0^{200} [30 - (0.1Q + 10)]\,dQ = \int_0^{200} [20 - 0.1Q]\,dQ = 2000$$

9.4 Mathematica examples

Calculating indefinite integrals with Mathematica is very straightforward. Suppose we wish to calculate the integral

$$\int 2x^3\,dx$$

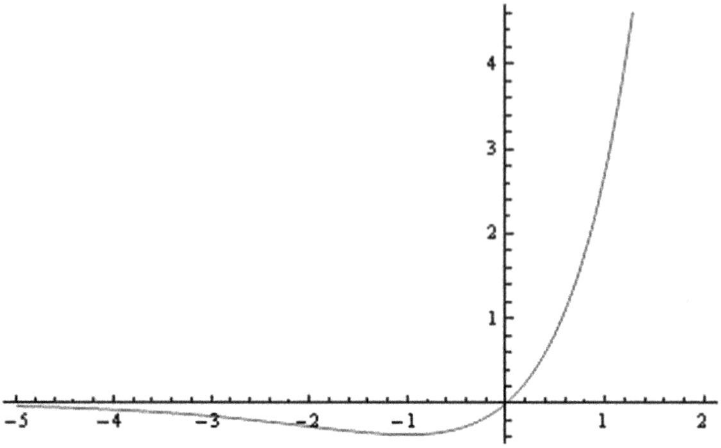

Figure 9.3

We use Mathematica code below to obtain the result.

```
>Integrate[2x ^3, x]
```

$$\frac{x^4}{2}$$

Mathematica handles more complicated integrals very well. For instance, to compute the integral

$$\int xe^x\, dx$$

we can use the code:

```
> Integrate[x * Exp[x], x]
```

$e^x(-1+x)$

It would help sometimes to graph the function we are about to integrate:

```
> Plot[x * Exp[x], {x, -5, 2}]
```

The graph is shown in Figure 9.3. We can compute the area under the graph of the function in Figure 9.3 over the interval [0, 1] as follows:

```
> Integrate[x * Exp[x], {x, 0, 1}]
```

1

PROBLEMS FOR SECTION 9.4

1. Compute the area under the graphs of the following functions over the indicated intervals.

   (a) $f(x) = x^2$ over $[-1, 1]$
   (b) $f(x) = x^3$ over $[0, 1]$
   (c) $f(x) = x^3$ over $[-1, 1]$
   (d) $f(x) = 2x + 5$ over $[2, 5]$
   (e) $f(x) = x^2 + x - 1$ over $[0, 2]$
   (f) $f(x) = (x - 1)^3$ over $[2, 4]$

   Explain why the area in part (c) is equal to zero.
2. Evaluate the following integrals.

   (a) $\int_0^5 2x\,dx$

   (b) $\int_0^3 (x^2 + x)\,dx$

   (c) $\int_2^4 t^2\,dt$

   (d) $\int_1^3 (x^3 - x)\,dx$

   (e) $\int_{-1}^0 (w^5 - w^2)\,dw$

   (f) $\int_{-3}^{-2} (z + 2z^2)\,dz$

3. Compute the area A bounded by the graphs of $f(x) = x^2$ and $g(x) = x$.
4. Compute the area A bounded by the following graphs.

   (a) $f_1(x) = x + 2$
   (b) $f_2(x) = x + 5$
   (c) $x = 2$
   (d) $x = 5$

Draw the area in question.

10 Mathematics of finance

In this chapter, we introduce the reader to one of the most important concepts in finance – the time value of money. The concept of the time value of money answers such questions as the following. If \$5000 is deposited in a savings account that earns interest at the annual rate of 12 percent, how much money will be in the account seven years from now? What is the present value of \$3000 due in five years if the interest rate is 7 percent compounded quarterly? Would you rather have a savings account that pays 6 percent compounded annually or 5 percent compounded continuously? We show how to compute the present and future values of a stream of cash flows and how to compare the future results of an investment with its present cost. Certain types of assets, such as annuities and perpetuities, have cash flows that follow a regular pattern. We also discuss annuities, perpetuities, mortgages, car loans, student loans and the effective rate of interest.

10.1 Simple interest

Suppose that you have loaned \$1000 to your friend, who will pay back the principal and interest in five years. For each year that he holds the money, you are to receive \$110 or 11 percent of the principal amount. How much money will you receive in five years? The answer is simple:

$$\$1000(1+0.11\cdot 5)=\$1550$$

In general, if you borrow \$1000 for t periods of time at simple interest, your interest bill would be

$$I=P_0rt$$

where P_0 is the principal amount, r is the interest rate and t is the number of periods the loan is held for. Further, the total payment due on a simple interest loan is

$$\text{Total payment due}=P_0(1+rt)$$

The simple interest method does not charge interest on interest and is not widely used in practice.

10.2 Compound interest

The vast majority of lenders require the borrower to pay compound interest on a loan. Further, most deposit accounts at banks and other financial institutions pay compound interest on the balance in the account. The future value FV of an asset earning compound interest is given by

$$\text{FV} = P_0(1+r)^t$$

where r is the interest rate and t is the number of periods (typically years) an asset is held for.

For example, suppose \$2000 is lent for eight years at 5 percent per year, compounded annually. Then, the lump sum the borrower must pay at the end of eight years is

$$\text{FV} = \$2000(1+0.05)^8 \approx \$2954.91$$

Thus, the compound interest on the loan is

$$\text{Compound interest} = \text{FV} - P_0 \approx \$2954.91 - \$2000 = \$954.91$$

Increased competition between financial institutions has forced many depositary institutions to compound interest more frequently than once per annum, with some depositary institutions compounding interest monthly or even daily. Now, let us say you hold an \$8000 deposit in a bank that offers 12 percent interest rate compounded monthly. At the end of five years your account balance will be

$$\text{FV} = \$8000\left(1+\frac{0.12}{12}\right)^{5\cdot 12} \approx \$14{,}533.57$$

The general formula for the future value of an asset that pays interest of r percent per annum, compounded n times per year, is given by

$$\text{FV} = P_0\left(1+\frac{r}{n}\right)^{tn}$$

To summarize, *future value* (FV) is the amount of money that an investment made today will become at some future date, given a certain fixed interest rate.

Example 10.1 Suppose you deposit \$2000 in a bank that offers a 5 percent interest rate on time deposits (compounded annually). In seven years the \$2000 will grow to

$$\text{FV} = \$2000(1+0.05)^7 \approx \$2814.20$$

In other words, the future value of \$2000 in seven years is approximately \$2814.20 if the interest rate is 5 percent.

Example 10.2 Let us say you hold an \$8000 deposit in an account that earns interest at the annual rate of 12 percent. How much money will there be in the account in seven years if the interest is compounded

(a) annually?
(b) quarterly?
(c) weekly?
(d) daily?

Solution:

(a) If the interest is compounded annually, the *FV* of the account in seven years is

$$\text{FV} = P_0\left(1+\frac{r}{n}\right)^{tn} = \$8000(1+0.12)^7 \approx \$17{,}685.45$$

(b) If the interest is compounded quarterly, the *FV* of the account in seven years is

$$\text{FV} = P_0\left(1+\frac{r}{n}\right)^{tn} = \$8000\left(1+\frac{0.12}{4}\right)^{7\cdot 4} \approx \$18{,}303.42$$

(c) If the interest is compounded weekly, the *FV* of the account in seven years is

$$\text{FV} = P_0\left(1+\frac{r}{n}\right)^{tn} = \$8000\left(1+\frac{0.12}{52}\right)^{7\cdot 52} \approx \$18{,}513.01$$

(d) If the interest is compounded daily, the *FV* of the account in seven years is

$$\text{FV} = P_0\left(1+\frac{r}{n}\right)^{tn} = \$8000\left(1+\frac{0.12}{365}\right)^{7\cdot 365} \approx \$18{,}528.38$$

Note that, the more often the interest is compounded, the larger is the future value.

EXAMPLE 10.3 How many years will it take for \$8000 in the previous example to triple, if interest is compounded once per annum?

Solution: The future value of the account after t years (where t does not have to be an integer) is

$$\text{FV} = P_0(1+r)^t$$

Thus, we need to solve the equation

$$P_0(1+r)^t = 3P_0 \tag{10.1}$$

for t. To find t we divide equation (10.1) by P_0 and take the natural logarithm of each side:

$$\ln(1+r)^t = \ln 3$$

It follows immediately that

$$t = \frac{\ln 3}{\ln(1+r)} = \frac{\ln 3}{\ln(1.12)} \approx 9.7 \text{ years}$$

Thus it will take approximately 9.7 years for the account to triple.

10.3 Continuous compounding

In the previous section we saw that interest can be compounded quarterly, monthly, weekly and even daily. In principle, there is no limit to how often we can compound interest. Consider the future value expression below:

$$\mathrm{FV} = P_0\left(1+\frac{r}{n}\right)^{tn}$$

To find what happens to FV as n tends to infinity, consider the following limit:

$$\lim_{n\to\infty}\left[P_0\left(1+\frac{r}{n}\right)^{tn}\right]$$

If we let $k = \frac{n}{r}$, then

$$\begin{aligned}
\lim_{n\to\infty} P_0\left(1+\frac{r}{n}\right)^{tn} &= \lim_{n\to\infty} P_0\left(1+\frac{1}{k}\right)^{krt}\\
&= P_0 \lim_{n\to\infty}\left(1+\frac{1}{k}\right)^{krt}\\
&= P_0 \lim_{k\to\infty}\left[\left(1+\frac{1}{k}\right)^{k}\right]^{rt}\\
&= P_0\left[\lim_{k\to\infty}\left[\left(1+\frac{1}{k}\right)^{k}\right]\right]^{rt} = P_0\, e^{rt}
\end{aligned}$$

Thus, under continuous compounding the future value of an asset is

$$\mathrm{FV} = P(t) = P_0\, e^{rt}$$

where t is usually measured in years.

EXAMPLE 10.4 Suppose that \$8000 is invested in an account that yields interest at an annual rate of 15 percent compounded continuously. What will the account balance be after five years?

Solution: The account balance after five years will be

$$\mathrm{FV} = P_0\, e^{rt} = \$8000 \cdot e^{0.15\cdot 5} \approx \$16{,}936.00$$

10.4 Effective annual rate

Suppose you are interested in opening a savings account with one of the banks in town. Bank N pays 8 percent interest, compounded semi-annually, on time deposits. Bank C pays 6 percent interest, compounded quarterly. In which bank should you open an account?

To compare the two offers, we shall use the concept of effective annual rate (EAR). EAR is defined as

$$\text{EAR} = \left(1 + \frac{r}{n}\right)^n - 1$$

where r is the quoted interest rate and n is the number of times that the interest is compounded per year.

For example, the EAR that Bank N pays is

$$\text{EAR}_{\text{BN}} = \left(1 + \frac{0.08}{2}\right)^2 - 1 \approx 0.0816 = 8.16 \text{ percent}$$

Similarly, the EAR that Bank C pays is

$$\text{EAR}_{\text{BC}} = \left(1 + \frac{0.06}{4}\right)^4 - 1 \approx 0.0614 = 6.14 \text{ percent}$$

Thus, you should choose Bank N to be your saving institution.

EXAMPLE 10.5 What is the EAR corresponding to a quoted annual interest rate of 9 percent with interest compounded

(a) monthly?
(b) quarterly?
(c) semi-annually?

Solution:

(a) The EAR with interest compounded monthly is

$$\text{EAR} = \left(1 + \frac{0.09}{12}\right)^{12} - 1 \approx 0.0938 = 9.38 \text{ percent}$$

(b) The EAR with interest compounded quarterly is

$$\text{EAR} = \left(1 + \frac{0.09}{4}\right)^4 - 1 \approx 0.0930 = 9.30 \text{ percent}$$

(c) The EAR with interest compounded semi-annually is

$$\text{EAR} = \left(1 + \frac{0.09}{2}\right)^2 - 1 \approx 0.0920 = 9.20 \text{ percent}$$

10.5 Present value

Given a choice between \$1000 today or \$1000 five years from now, a rational person will choose \$1000 today. The choice is obvious since one can invest the \$1000 and earn some interest r in the meantime. If the interest rate $r = 0.06$ and the interest is paid annually, then after five years the original amount of \$1000 will have grown to $\$1000(1+0.06)^5 = \$1338.23$.

A rational person would be indifferent between receiving \$1000 today or \$1338.23 five years from now, given that the interest rate r is fixed at 6 percent. To compare the value of money today with the value of money in the future, we shall introduce the concept of present value. The *present value* (PV) of a *single future cash flow* (CF) that an investor expects to receive t years from now is

$$\mathrm{PV} = \frac{\mathrm{CF}}{(1+r)^t}$$

where r is the risk-adjusted (or required) interest rate. The expression

$$\frac{1}{(1+r)^t}$$

is known as the *discount factor.*

If interest is compounded continuously, then the discount factor is e^{-rt} and the *present value* (PV) of a single future cash flow (CF) is

$$\mathrm{PV} = \frac{\mathrm{CF}}{e^{rt}}$$

As you would expect, the present value of an asset with continuous compounding is smaller than the present value with annual compounding.

EXAMPLE 10.6 What is the present value of \$20,000 to be paid after seven years if the required interest rate is 8 percent, compounded

(a) annually?
(b) monthly?
(c) continuously?

Solution:

(a) The present value is

$$\mathrm{PV}_{\text{annual}} = \frac{20{,}000}{(1+0.08)^7} = \$11{,}669.80$$

(b) The present value is

$$\mathrm{PV}_{\text{monthly}} = \frac{20{,}000}{(1+\frac{0.08}{12})^{7\cdot 12}} = \$11{,}445.43$$

(c) The present value is

$$\text{PV}_{\text{continuously}} = \frac{20,000}{e^{0.08 \cdot 7}} = \$11,424.18$$

Now, suppose that four successive annual payments are to be received, in the amount of CF_1 after the third year, then CF_2 after the fourth year, then CF_3 after the fifth year, and CF_4 after the sixth year. How do we compute the present value of the stream of income in question?

The present value of the stream of income is computed as the sum of present values of individual cash flows:

$$\text{PV} = \frac{\text{CF}_1}{(1+r)^3} + \frac{\text{CF}_2}{(1+r)^4} + \frac{\text{CF}_3}{(1+r)^5} + \frac{\text{CF}_4}{(1+r)^6}$$

The stream of income discussed above is similar to an annuity. An annuity is defined as a stream of N equal cash flows paid at regular intervals. The cash flows of an annuity can be represented on a timeline in the following manner.

The first payment of an annuity takes place on date 1, one period from today; the second payment takes place on date 2, two periods from now, and so on. Thus, the present value of the annuity in question is:

$$\text{PV} = \frac{\text{CF}}{(1+r)^1} + \frac{\text{CF}}{(1+r)^2} + \cdots + \frac{\text{CF}}{(1+r)^N}$$

Using the summation formula for finite geometric series we obtain

$$\text{PV} = \frac{\text{CF}}{(1+r)^1} + \frac{\text{CF}}{(1+r)^2} + \cdots + \frac{\text{CF}}{(1+r)^N} = \frac{\text{CF}}{r}\left[1 - \frac{1}{(1+r)^N}\right]$$

EXAMPLE 10.7 Suppose you are the lucky winner of the \$10 million national lottery. You can receive your money either as

(a) \$5 million paid today or as
(b) 20 payments of \$500,000 per year (starting one year from now).

Given that the interest rate is 5 percent, which option is better?

Solution: The present value of \$5 million paid today is exactly \$5 million. The present value of the second option is

$$\text{PV}_2 = \frac{\$500{,}000}{0.05}\left[1 - \frac{1}{(1+0.05)^{20}}\right] \approx \$6,231,105$$

Therefore, option (b) is more valuable.

EXAMPLE 10.8 Suppose you are given a choice between two annuities, say A and B. Annuity A pays \$1000 per year for ten years and annuity B pays \$2000 per year for five years. Which annuity would you prefer if the interest rate is 5 percent.

Solution: As in the previous example, we need to compare the present values of the two annuities. Since

$$\mathrm{PV_A} = \frac{\$1000}{0.05}\left[1 - \frac{1}{(1+0.05)^{10}}\right] \approx \$7721.73$$

$$\mathrm{PV_B} = \frac{\$2000}{0.05}\left[1 - \frac{1}{(1+0.05)^{5}}\right] \approx \$8658.95$$

we conclude that annuity B is a better option. Note, that $\mathrm{PV_B}$ is much larger than $\mathrm{PV_A}$.

A *perpetuity* is a stream of equal cash payments that occur at fixed intervals and last forever. A perpetuity is essentially an annuity that has no definite end. There are few actual perpetuities in existence. (The British government has issued perpetuities in the past. These perpetuities are known as *perpetual bonds* or *consols*.) However, a number of types of investments are effectively perpetuities (e.g. real estate or preferred stock) and can certainly be valuated just as perpetuities.

The present value of a perpetuity with payments CF and interest rate r is given by

$$\mathrm{PV} = \frac{\mathrm{CF}}{(1+r)^1} + \frac{\mathrm{CF}}{(1+r)^2} + \cdots = \frac{\mathrm{CF}}{r}$$

You can, in effect, create a perpetuity by buying a new house for \$100,000 and renting it for \$6000 a year. In this scenario, you will be effectively purchasing a perpetuity that pays \$6000 a year forever. Alternatively, you can invest the \$100,000 in a bank that pays 6 percent per year and collect interest forever.

EXAMPLE 10.9 Suppose the British government has issued a consol bond that pays £10,000 a year forever. What is the value of the bond right after a payment is made if the interest rate is 5 percent?

Solution: The present value of the consol bond is

$$\mathrm{PV_{consol}} = \frac{\mathrm{CF}}{(1+r)^1} + \frac{\mathrm{CF}}{(1+r)^2} + \cdots = \frac{\mathrm{CF}}{r} = \frac{£10{,}000}{0.05} = £200{,}000$$

The *future value* (FV) of a stream of income is computed as the sum of future values of individual cash flows. The future value of a stream of income can be computed at any point in time. However, typically the future value is computed by compounding each payment to the end of the stream of income and then adding the future values, as shown below:

$$\begin{aligned}\mathrm{FV}_n &= \frac{\mathrm{CF}_1}{(1+r)^{k-1}} + \frac{\mathrm{CF}_2}{(1+r)^{k-2}} + \cdots + \frac{\mathrm{CF}_k}{(1+r)^{k-k}} \\ &= \sum_{t=1}^{k} \frac{\mathrm{CF}_k}{(1+r)^{k-t}}\end{aligned}$$

We can use the formula above to find the future value of an ordinary annuity. Let us say that you have an annuity that pays $500 per year for ten years at 12 percent. What is the future value of such an annuity after ten years?

The future value of the ordinary annuity (FVOA) in question is

$$\begin{aligned}
\mathrm{FVOA}_{10} &= \frac{\mathrm{CF}}{(1+r)^{10-1}} + \frac{\mathrm{CF}}{(1+r)^{10-2}} + \cdots + \frac{\mathrm{CF}}{(1+r)^{0}} \\
&= \frac{\mathrm{CF}}{(1+r)^{9}} + \frac{\mathrm{CF}}{(1+r)^{8}} + \cdots + \frac{\mathrm{CF}}{(1+r)^{0}} \\
&= \$500\left(\frac{1}{(1+0.12)^{9}} + \frac{1}{(1+0.12)^{8}} + \cdots + \frac{1}{(1+0.12)^{0}}\right) \\
&= \$500(17.5487) \approx \$8774.37
\end{aligned}$$

Calculating the future value of an ordinary annuity using the formula above may be tedious. Luckily, the equation above can be simplified and the future value of an ordinary annuity can be found using the formula below:

$$\mathrm{FVOA}_t = \mathrm{CF}\left[\frac{(1+r)^t - 1}{r}\right] \tag{10.2}$$

Using equation (10.2), the future value of an annuity that pays $500 per year for ten years at 12 percent is computed as

$$\mathrm{FVOA}_{10} = \$500\left[\frac{(1+0.12)^{10} - 1}{0.12}\right] \approx \$8774.37$$

Example 10.10 What is the future value of $3000 in five years if the interest rate is

(a) 7 percent compounded annually?
(b) 18 percent compounded annually?
(c) 5 percent compounded continuously?

Solution:

(a) The future value FV_1 is

$$\mathrm{FV}_1 = \$3000(1+0.07)^5 \approx \$4207.66$$

(b) The future value FV_2 is

$$\mathrm{FV}_2 = \$3000(1+0.18)^5 \approx \$6863.27$$

(c) The future value FV_3 is

$$\mathrm{FV}_3 = \$3000\, e^{0.05 \cdot 5} \approx \$3852.07$$

PROBLEMS FOR SECTION 10.5

1. Compute the present value of $1,000,000 that is due in 20 years if the interest rate is 5 percent per year compounded annually.
2. Compute the present value of an annuity that pays $60,000 at the end of each year for 30 years if the interest rate is 5 percent compounded annually.
3. Compute the present value of $2,000,000 that is due in 50 years if the interest rate is 5 percent compounded continuously.
4. Compute the present value of a series of payments of $5000 at the end of each year in perpetuity if the interest rate is

   (a) 2 percent per year,
   (b) 5 percent per year,
   (c) 12 percent per year.

   Assume that interest is compounded annually.
5. An amount of $10,000 earns interest at 8 percent per year. How long will it take for the $10,000 to quadruple?
6. Calculate the future value of $5000 in

   (a) five years if the interest rate is 6 percent per year compounded annually,
   (b) six years if the interest rate is 7 percent per year compounded annually,
   (c) seven years if the interest rate is 8 percent per year compounded continuously.

7. Find the present value of the following ordinary annuities:

   (a) $200 per year for three years at 4 percent;
   (b) $300 per year for four years at 5 percent;
   (c) $400 per year for five years at 8 percent;
   (d) $500 per year for seven years at 12 percent.

8. At 18 percent compounded monthly, how many months are necessary for

   (a) $5000 to grow to $8000?
   (b) $900 to grow to $2700?
   (c) $1200 to grow to $1500?
   (d) $11,500 to grow to $15,350?

9. What is the present value of $5000

   (a) paid at the end of each year for 18 years?
   (b) paid at the begining of each year for 18 years?

10. Your broker offers you an annuity that will pay $20,000 a year for eight years starting at the end of this year. If you require 12 percent return per annum, what is the maximum amount of money you should pay for the annuity?

10.6 Car loans and mortgages

Car loans, mortgages and student loans are paid off in installments over time. Loans that are paid off in equal periodic amounts are known as *amortized loans* and, as you might have realized already, car loans, student loans and mortgages are all examples of amortized loans. Typically, mortgages and cars loans are paid off on a monthly basis.

Suppose an investor borrows $50,000 from a bank, and he is supposed to pay it off in five equal installments at the end of each of the next five years, with interest at 7 percent compounded annually. Let us determine the annual payment that the investor has to make to the bank. To find the annual payment, we need to realize that for $50,000 the bank is effectively purchasing an annuity of CF dollars per year for five years. Therefore, we have

$$\$50{,}000 = \left[\sum_{t=1}^{5} \frac{\text{CF}}{(1+r)^t}\right] = \text{CF}\left[\frac{1 - \frac{1}{1.07^5}}{0.07}\right] = \text{CF}(4.1002)$$

Now, we can solve for CF as follows:

$$\text{CF} = \frac{\$50{,}000}{4.1002} \approx \$12{,}194.52$$

Thus the investor must pay the bank $12,194.52 at the end of each of the next five years, given that the interest rate is 7 percent.

EXAMPLE 10.11 Suppose a firm has just purchased a house using a $100,000 fixed-rate, ten-year mortgage that has an annual interest rate of 3 percent. Determine the dollar amount of the firm's mortgage payment given that the firm is supposed to pay it off on a monthly basis.

Solution: We know that the present value of the mortgage is $100,000, the monthly interest rate is 0.25 percent and there are supposed to be 120 payments (PMT). Thus, we know that

$$\begin{aligned} \$100{,}000 &= \sum_{t=1}^{120} \frac{\text{PMT}}{(1+r)^t} \\ &= \text{PMT}\left[\frac{1 - \frac{1}{1.0025^{120}}}{0.0025}\right] = \text{PMT}(103.5617) \end{aligned}$$

Therefore, the firm's mortgage payment will be

$$\text{PMT} = \frac{\$100{,}000}{103.5617} = \$965.61$$

Incidentally, in Example 10.11, the total amount of money that the firm will pay for the house is

$$\text{TotPayment} = \$965.61 \cdot 120 = \$115{,}873.29$$

Out of that amount, $15,873, or about 16 percent of the total amount borrowed, will be used to cover the interest charges. Now, let us say that the firm in question wants to purchase another house using a $100,000 fixed-rate, 30-year mortgage that has an annual interest rate of 3 percent. Then, the firm's second mortgage payment PMT_2 will be

$$PMT_2 = \frac{\$100{,}000}{\left[\dfrac{1 - \dfrac{1}{1.0025^{360}}}{0.0025}\right]} = \$421.60$$

and the total amount of money that the firm will pay for the second house will be

$$\text{TotPayment}_2 = \$421.60 \cdot 360 = \$151{,}777.45$$

Note that in this case interest charges exceed 51 percent of the total amount borrowed. Many firms and individuals may not realize the financial upside of having a fixed 10- or 15-year mortgage. Besides paying less total interest, they typically have significantly lower interest rates than 30-year fixed mortgages.

EXAMPLE 10.12 Jason is considering purchasing a new car and his local dealer's best price is $18,000. The dealer offers 0 percent financing for 60 months, or $2000 rebate and 9 percent financing for 60 months. Assume that Jason is planning to sell his new car in exactly three years. Which of the two financing options would you recommend to Jason?

Solution: If Jason choses the first option, his monthly payment PMT_1 will be $300 for 60 months and his payoff at the end of the three years will be

$$\text{Payoff}_1 = \$300 \cdot 24 = \$7200$$

Should Jason choose the second option, his purchasing price will be $16,000 and his monthly payment PMT_2 therefore will be

$$PMT = \frac{\$16{,}000}{\left[\dfrac{1 - \dfrac{1}{1.0075^{60}}}{0.0075}\right]} \approx \$332.13$$

Jason's payoff at the end of three years can be found as follows:

$$PV = \frac{\$332.13}{0.0075}\left[1 - \frac{1}{(1+0.0075)^{24}}\right] = \$7270.04$$

So, both Jason's payments and payoff are lower in the first case. Clearly, Jason should take advantage of the 0 percent financing option.

PROBLEMS FOR SECTION 10.6

1. A car dealership wants to purchase another building and has the following two choices:

   (a) Pay \$20,000 per month for nine years, with the first payment due at the end of the first month.

   (b) Pay \$12,000 per month for 15 years, with the first payment due at the beginning of the first month.

   Determine which option is the better if the quoted interest rate is 12 percent.

2. A firm borrows \$50,000 from a credit union at the beginning of a year, and is supposed to pay it off in 120 equal installments at the end of each month. Assume that the quoted interest rate is 9 percent.

   (a) Find the monthly payment and the total amount of money that the firm will pay to the credit union.

   (b) Find the total amount of money that the firm will pay to the credit union if the interest rate is 12 percent.

3. Suppose you have an outstanding loan that requires 36 monthly payments of \$400 and the next payment is due one month from now. What will your payoff on the loan be exactly one year from now?

4. Your broker offers to sell you a time share for \$15,000 that will pay \$3000 per year for seven years. If you buy the time share, what rate of interest will you be earning?

11 Complex numbers

In this chapter we introduce an extension of real numbers called complex numbers. At first, it is difficult to understand the real-life importance of such an apparently abstract concept as complex numbers. However, there are deep mathematical reasons why complex numbers inevitably appear in many applications. For example, matrices with *real* entries can have *complex eigenvalues*. The behavior of such well-known functions as *sine* and *cosine* is properly understood only when we introduce complex numbers. Most importantly, one of the key questions in time-series analysis deals with the stability of the system. This question is reduced to finding eigenvalues of a certain matrix, which describe the ongoing autoregressive process.

Solving difference or differential equations is greatly simplified by using complex numbers. Further, solving quadratic equations over complex numbers is simpler, as we are guaranteed to have solutions. Our goal is to present all of the necessary mathematical facts about complex numbers to prepare the reader for textbooks in time-series analysis and other areas that frequently deal with complex numbers.

11.1 The set of complex numbers

The set complex numbers $\mathbb{C}$ is an extension of the set of real numbers $\mathbb{R}$. The set of complex numbers contains all roots of quadratic equations. Let i be such that

$$i^2 = -1$$

then the set of complex numbers can be constructed as follows:

$$\mathbb{C} = \{a + bi \mid a, b \in \mathbb{R}\}$$

Complex numbers can be represented graphically by associating $z = a + bi$ with the point (a, b) on the complex plane. Figure 11.1 demonstrates how to locate a complex number on the complex plane.

Now, let $z = a + bi$ be a complex number. Then a is the real part of z and b is the imaginary part of z and we shall write

$$a = \text{Re(z)}$$
$$b = \text{Im(z)}$$

to indicate that fact.

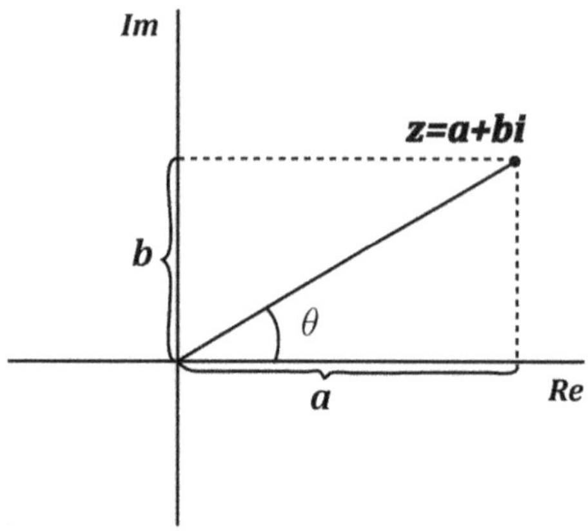

Figure 11.1

Note the following points.

(a) Instead of $0+bi$, we shall write bi. Complex numbers of this form are known as purely imaginary numbers.
(b) Instead of $a+0i$, we shall write a.
(c) Instead of $1i$, we shall write i.

Since the real part of complex numbers is represented on the x-axis we shall refer to the x-axis as the real axis. For similar reasons, we shall refer to the y-axis as the imaginary axis. It turns out that all real numbers may be written in the form $z=a+bi$. For numbers that are real, the imaginary part b merely equals zero. For instance, we may write 3 as $3+0i$. Complex numbers that have no real part are known as purely imaginary. The purely imaginary numbers are located on the y-axis.

11.1.1 Operations with complex numbers

Two complex numbers $z_1=a+bi$ and $z_2=c+di$ are equal if and only if $a=c$ and $b=d$. In other words,

$$a+bi=c+di \quad \Leftrightarrow \quad a=c \text{ and } b=d$$

The sum and the difference of complex numbers is defined as follows:

$$(a+bi)+(c+di)=(a+c)+(b+d)i$$

$$(a+bi)-(c+di)=(a-c)+(b-d)i$$

EXAMPLE 11.1

(a) $(2+3i)+(4+5i)=6+8i$
(b) $(5+7i)+(2-3i)=7+4i$
(c) $(7+9i)-(4+5i)=3+4i$

The product of two complex numbers is given by:

$$(a+bi)(c+di)=(ac-bd)+(ad+bc)i$$

EXAMPLE 11.2

(a) $(2+3i)(4+5i)=(8-15)+(10+12)i=-7+22i$
(b) $(5+7i)(2+3i)=(10-21)+(15+14)i=-11+29i$
(c) $(2+3i)(4+5i)=(8-15)+(10+12)i=-7+22i$

Now we shall define the conjugate of $a+bi$ as

$$\overline{a+bi}=a-bi$$

(see Figure 11.2) and denote it with a bar.

EXAMPLE 11.3

(a) $\overline{2+3i}=2-3i$

(b) $\overline{5-7i}=5+7i$

(c) $\overline{8i}=-8i$

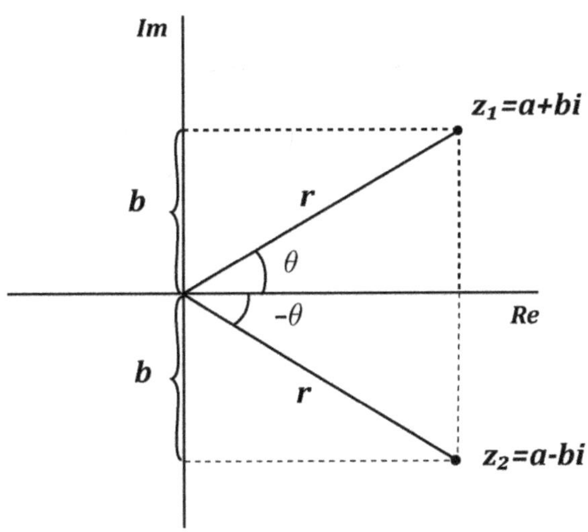

Figure 11.2

The modulus or the absolute value of a complex number $z = a + bi$ is

$$|z| = \sqrt{a^2 + b^2}$$

and sometimes written as mod(z). If a point A represents $z = a + bi$ on the complex plane, then the distance from the origin O to the point A is the modulus of z

$$|z| = |OA|$$

Note that compex conjugates have the same modulus:

$$|\overline{z}| = |z|$$

EXAMPLE 11.4

(a) $|3 + 4i| = 5$
(b) $|3 - 4i| = 5$
(c) $|1 - 2i| = \sqrt{5}$
(d) $|2 + \sqrt{12}i| = 4$

Using complex numbers we can solve any quadratic equation. Recall that the solution for $ax^2 + bx + c = 0$ is given by

$$x_{1,2} = \frac{-b \pm \sqrt{D}}{2a}$$

where

$$D = b^2 - 4ac$$

We can use the formula over the complex numbers even when the discriminant D is negative.

EXAMPLE 11.5 Solve

$$z^2 + z + 2 = 0$$

Solution:

$$\begin{aligned} z_{1,2} &= \frac{-1 \pm \sqrt{1 - 4 \cdot 2}}{2} \\ &= \frac{-1 \pm \sqrt{-7}}{2} = \frac{-1 \pm i\sqrt{7}}{2} \\ &= -\frac{1}{2} \pm \frac{\sqrt{7}}{2} i \end{aligned}$$

EXAMPLE 11.6 Prove that $\overline{z_1 z_2} = \overline{z_1} \cdot \overline{z_2}$.

Solution: Let $z_1 = a + bi$ and $z_2 = c + di$. Then

$$\begin{aligned}
\overline{z_1 z_2} &= \overline{(a+bi)(c+di)} = \overline{ac + bic + adi + bdi^2} \\
&= \overline{(ac-bd) + i(bc+ad)} = (ac-bd) - i(bc+ad) \\
&= (a-bi)(c-di) = \overline{(a+bi)(c+di)} \\
&= \overline{z_1} \cdot \overline{z_2}
\end{aligned}$$

EXAMPLE 11.7 Show that $\overline{z_1 + z_2} = \overline{z_1} + \overline{z_2}$.

Solution: Let $z_1 = a + bi$ and $z_2 = c + di$. Then

$$\begin{aligned}
\overline{z_1 + z_2} &= \overline{a + bi + c + di} = \overline{a + c + (b+d)i} \\
&= (a+c) - (b+d)i = (a-bi) + (c-di) \\
&= \overline{z_1} + \overline{z_2}
\end{aligned}$$

11.2 Polar and trigonometric form of complex numbers

So far we have discussed the so-called Cartesian representation of the complex numbers. In essence, we have assigned to each complex number a point on the Cartesian (or complex) plane. However, complex numbers have also a representation in polar coordinates. A complex number $z = a + bi$ is assigned coordinates (r, θ) in the polar system, where

$$r = |z|$$

and θ is such that

$$\sin\theta = \frac{b}{r} \quad \text{and} \quad \cos\theta = \frac{a}{r}$$

as shown in Figure 11.3.

The pair (r, θ) specifies a unique point on the complex plane. However, a given point on the complex plane does not have a unique polar representation. In fact, a point $z = (r, \theta)$ has infinitely many polar representations of the form $z = (r, \theta + 2\pi n)$, with $n \in \mathbb{Z}$. A number in the form $(\theta + 2\pi n)$ is called *an argument* of z. The argument of z lying in the range $(-\pi, \pi]$ is referred to as the *principal argument* of z or Arg(z). For instance, the principal argument of $2i$ is $\frac{\pi}{2}$ and the principal argument of $(2i + 2)$ is $\frac{\pi}{4}$

$$\begin{aligned}
\text{Arg}(2i) &= \frac{\pi}{2} \\
\text{Arg}(2i+2) &= \frac{\pi}{4}
\end{aligned}$$

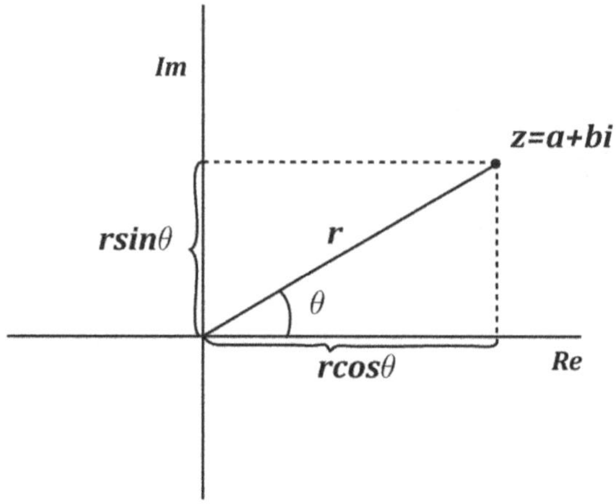

Figure 11.3

A complex number z can also be written in a trigonometric form:

$$z = r(\cos\theta + i\sin\theta) \tag{11.1}$$

The trigonometric form of complex numbers is used extensively in time-series analysis.

EXAMPLE 11.8 Find the trigonometric form of $z = 2 + 2i$.

Solution:

$$|z| = \sqrt{8}$$

$$\cos\theta = \frac{\sqrt{2}}{2}$$

$$\sin\theta = \frac{\sqrt{2}}{2}$$

It follows that

$$\theta = \mathrm{Arg}(z) = \frac{\pi}{4}$$

Hence

$$z = \sqrt{8}\left[\cos\left(\frac{\pi}{4}\right) + i\sin\left(\frac{\pi}{4}\right)\right].$$

Complex numbers can also be written in an exponential form. Euler's formula from complex analysis states that

$$e^{i\theta} = \cos\theta + i\sin\theta \tag{11.2}$$

and

$$e^{-i\theta} = \cos\theta - i\sin\theta \tag{11.3}$$

Reflecting equations (11.2) and (11.3) on (11.1) gives the exponential representation of complex numbers z and $\overline{z}$:

$$z = r(\cos\theta + i\sin\theta) = r\,e^{i\theta}$$
$$\overline{z} = r(\cos\theta - i\sin\theta) = r\,e^{-i\theta}$$

Now, let $z_1 = a + bi$ and $z_2 = c + di$. Then

$$\begin{aligned} z_1 \cdot z_2 &= (a+bi)(c+di) \\ &= r_1(\cos\theta_1 + i\sin\theta_1)\cdot r_2(\cos\theta_2 + i\sin\theta_2) \\ &= r_1r_2(\cos\theta_1\cos\theta_2 + i\cos\theta_1\sin\theta_2 + i\sin\theta_1\cos\theta_2 - i\sin\theta_1\sin\theta_2) \\ &= r_1r_2[\cos(\theta_1+\theta_2) + i\sin(\theta_1+\theta_2)] \end{aligned}$$

The product of z_1 and z_2 is represented graphically in Figure 11.4. Note that when we multiply complex numbers, we multiply moduli and add arguments.

In particular, if $z_1 = z_2 = a + bi$ we have

$$z^2 = (a+bi)^2 = r^2[\cos(2\theta) + i\sin(2\theta)]$$
$$z^3 = (a+bi)^3 = r^3[\cos(3\theta) + i\sin(3\theta)]$$

See Figure 11.5 for the graphical representation of z^2 and z^3.

To find the quotient of two complex numbers $a + bi$ and $c + di$ we use the complex conjugate of the denominator as follows:

$$\begin{aligned} \frac{a+bi}{c+di} &= \frac{(a+bi)\overline{(c+di)}}{(c+di)\overline{(c+di)}} \\ &= \frac{(a+bi)(c-di)}{(c+di)(c-di)} = \frac{(ac+bd)+i(bc-ad)}{c^2+d^2} \end{aligned}$$

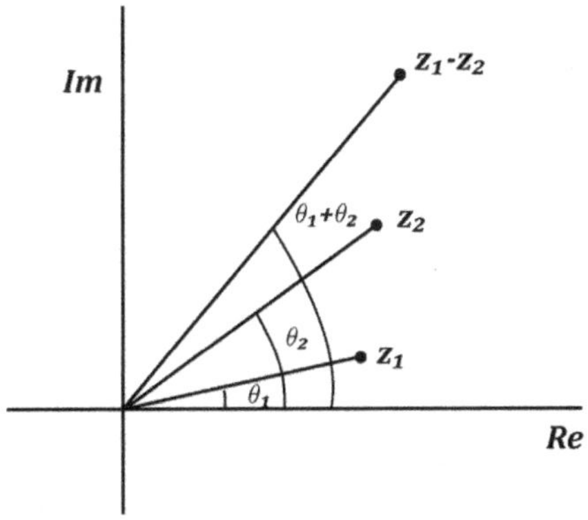

Figure 11.4

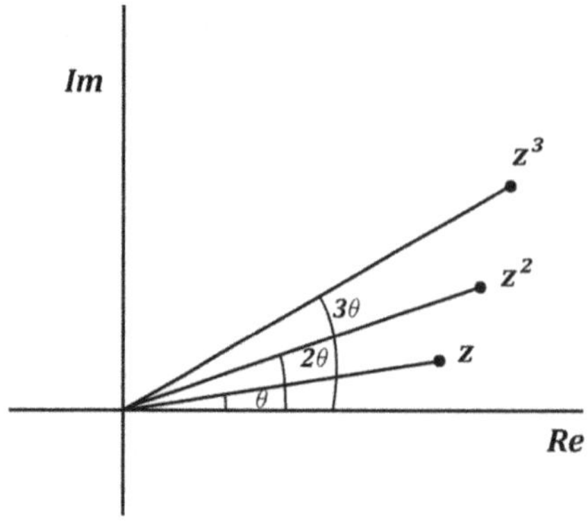

Figure 11.5

EXAMPLE 11.9 Find the quotient of complex numbers $z_1 = 2 + 3i$ and $z_2 = 5 + 6i$

Solution:

$$\frac{2+3i}{5+6i} = \frac{(2+3i)\overline{(5+6i)}}{(5+6i)\overline{(5+6i)}} = \frac{(2+3i)(5-6i)}{(5+6i)(5-6i)}$$
$$= \frac{28+3i}{25+36} = \frac{28}{61} + \frac{1}{12}i$$

One of the very important results in complex analysis is known as De Moivre's theorem, named after the English mathematician of French origin, Abraham de Moivre (1667–1754). The theorem states that if

$$z = r\,(\cos\theta + i\sin\theta)$$

then

$$z^n = r^n(\cos n\theta + i\sin n\theta) \quad \forall\, n \in \mathbb{N}$$

EXAMPLE 11.10 Find $(2+2i)^{10}$ using De Moivre's theorem.

Solution: Since $|z| = \sqrt{8}$ and $\mathrm{Arg}(z) = \dfrac{\pi}{4}$ then

$$z^{10} = \left[\sqrt{8}\left(\cos\frac{\pi}{4} + i\sin\frac{\pi}{4}\right)\right]^{10} = 8^5\left(\cos\frac{5\pi}{2} + i\sin\frac{5\pi}{2}\right)$$

11.3 Mathematica examples

Consider two complex numbers $z_1 = 2 + 2i$ and $z_2 = 3 + 3i$. Using the following Mathematica code we can find the sum, difference, product and quotient of the two complex numbers.

>**(2 + 2I) + (3 + 3I)**
>**(2 + 2I) − (3 + 3I)**
>**(2 + 2I) ∗ (3 + 3I)**
>**(2 + 2I)/(3 + 3I)**

The code above produces the following output:

```
5+5I

-1-I

12I
```

$\frac{2}{3}$

We can find the principal arguments of the two complex numbers.

>**Arg[2 + 2I]**
>**Arg[3 + 3I]**

$\frac{\pi}{4}$

$\frac{\pi}{4}$

As you might have expected, the principal arguments of the two complex numbers above are equal: they both are $\frac{\pi}{4}$. This suggests that the principal argument of their product should be equal to $\frac{\pi}{2}$. We can verify that using Mathematica.

>**Arg[(2 + 2I) ∗ (3 + 3I)]**

$\frac{\pi}{2}$

Mathematica can help to identify the real and imaginary parts of a complex number.

>**Re[2 + 3I]**
>**Im[2 + 3I]**

```
2

3
```

It can also compute the absolute value and the conjugate of a number.

>**Abs[2 + 3I]**
>**Conjugate[2 + 3I]**

$\sqrt{13}$

```
2 - 3I
```

PROBLEMS FOR SECTION 11.3

1. Evaluate the expressions below and write your answer in the form $a + bi$.

   (a) $(5 - 6i) + (3 + 2i)$

   (b) $(2 + 5i)(4 - i)$

(c) $\overline{12+7i}$

(d) $\dfrac{1+4i}{3+2i}$

(e) $\dfrac{1}{1+i}$

(f) i^3

(g) $\sqrt{-25}$

(h) $(4-\frac{1}{2}i)-(9+\frac{5}{2}i)$

(i) $(1-2i)(8-3i)$

(j) $\overline{2i(\frac{1}{2}-i)}$

(k) $\dfrac{3+2i}{1-4i}$

(l) $\dfrac{3}{4-3i}$

(m) i^{100}

(n) $\sqrt{-3}\sqrt{-12}$

2. Prove the following properties of complex numbers.

(a) $\overline{z+w}=\overline{z}+\overline{w}$

(b) $\overline{zw}=\overline{z}\cdot\overline{w}$

(c) $\overline{z^n}=\overline{z}^n$

3. Find the complex conjugate and the modulus of the following.

(a) $z_1=12-15i$

(b) $z_2=-4i$

(c) $z_3=-1+2\sqrt{2}i$

4. Find all solutions of the following equations.

(a) $4x^2+9=0$

(b) $x^4=1$

(c) $x^2+2x+5=0$

(d) $2x^2-2x+1=0$

(e) $z^2+z+2=0$

(f) $z^2+\frac{1}{2}z+\frac{1}{4}=0$

5. Find the principal argument of the following.

(a) $z_1=-3$

(b) $z_2=1-i$

(c) $z_3=3+3i$

(d) $z_4=8i$

6. Find the principal argument of the following.
   (a) $z_1 = 5$
   (b) $z_2 = -7$
   (c) $z_3 = -1 + i$
   (d) $z_4 = -3 - 3i$
7. Find the indicated power for each of the following using De Moivre's theorem.
   (a) $(1+i)^{20}$
   (b) $(2\sqrt{3}+2i)^5$
   (c) $(1-\sqrt{3}i)^5$
   (d) $(1-i)^8$
8. Use De Moivre's theorem with $n = 3$ to express $\cos 3\theta$ and $\sin 3\theta$ in terms of $\cos\theta$ and $\sin\theta$.
9. Write each of the following numbers in the form $(\cos\theta + i\sin\theta)$ or $(\cos\theta - i\sin\theta)$.
   (a) $e^{\frac{i\pi}{2}}$
   (b) $e^{2\pi i}$
   (c) $e^{\frac{i\pi}{3}}$
   (d) $e^{-i\pi}$
   (e) $e^{i\pi}$
   (f) e^{-i}
10. Use Mathematica to compute the principal arguments, absolute values and conjugates of the following.
    (a) $3+3i$
    (b) $2-5i$
    (c) $7+11i$
    (d) $12i$

12 Difference and differential equations

Many economic processes demonstrate dynamic behavior. However, in some of these cases, the underlying system is guided by rather simple relationships between quantities and their rates of change. Whenever we have a relationship between functions and their derivatives or rates of change, we essentially deal with differential or difference equations. For functions of continuous variables, we consider differential equations. For quantities dependent on a discrete variable, we talk about difference equations.

Consider an economic process developing in time. Most likely the functions involved in the process will depend on their previous values. For instance, the current supply depends on the demand we observed a week, a month or a year ago. Whenever we have a relationship involving quantities at time t and their previous values at times $t-1$, $t-2$, and so on, we implicitly deal with a mathematical structure known as a difference equation. This chapter begins with the mathematical theory of such equations, considers some applications of difference equations to economic problems, and closes with differential equations.

12.1 Difference equations

12.1.1 Linear first-order difference equations

A linear first-order difference equation may be written as follows:

$$A(t)y_{t+1} + B(t)y_t = C(t) \quad (t = 0, 1, 2, \ldots) \tag{12.1}$$

where

$$A(t) \neq 0$$

For example, the equation

$$3^t y_{t+1} + 2^t y_t = 5^{t+2} \tag{12.2}$$

is a linear first-order difference equation.

Alternatively, the first-order difference equation in (12.1) may be written in the form:

$$y_{t+1} + \frac{B(t)}{A(t)}y_t = \frac{C(t)}{A(t)} \tag{12.3}$$

If $\dfrac{B(t)}{A(t)}$ and $\dfrac{C(t)}{A(t)}$ are constants, then we can write equation (12.3) as follows:

$$y_{t+1} + by_t = c$$

where b and c are some constants. The difference equation

$$y_{t+1} + by_t = c \tag{12.4}$$

is known as a first-order difference equation with constant coefficients. If $c = 0$, equation (12.4) is referred to as a homogeneous first-order difference equation.

Letting $b = -a$ allows us to rewrite equation (12.4) in the following form:

$$y_{t+1} = ay_t + c \tag{12.5}$$

We can solve equation (12.4) by recursive substitution as follows:

$$\begin{aligned}
y_1 &= ay_0 + c \\
y_2 &= ay_1 + c = a(ay_0 + c) + c = a^2 y_0 + ac + c \\
y_3 &= ay_2 + c = a(a^2 y_0 + ac + c) + c = a^3 y_0 + a^2 c + ac + c \\
y_4 &= ay_3 + c = a^4 y_0 + a^3 c + a^2 c + ac + c \\
&\vdots \\
y_t &= a^t y_0 + c(1 + a + a^2 + \cdots + a^{t-1})
\end{aligned}$$

If $a = 1$, then the solution of $y_{t+1} = ay_t + c$ is given by

$$y_t = y_0 + ct \quad \text{where } a = 1 \tag{12.6}$$

If $a \neq 1$, then

$$1 + a + a^2 + \cdots + a^{t-1} = \frac{1 - a^t}{1 - a}$$

and the solution to (12.5) is given by

$$y_t = a^t y_0 + c\left(\frac{1 - a^t}{1 - a}\right) \quad \text{where } a \neq 1$$

or alternatively by

$$y_t = a^t\left(y_0 - \frac{c}{1 - a}\right) + \frac{c}{1 - a} \quad \text{where } a \neq 1 \tag{12.7}$$

The constant $y^* = \dfrac{c}{1 - a}$ is known as the equilibrium state of the first-order difference equation (12.5).

The sequence defined in (12.7) represents a *general solution* for the first-order difference equation

$$y_{t+1} = ay_t + c \quad \text{with } a \neq 1$$

A particular solution of the difference equation can be obtained from the general solution by giving a particular value to the constant y_0. The constant y_0 is normally referred to as the *initial value* of the difference equation.

EXAMPLE 12.1 Solve the first-order difference equation

$$8y_{t+1} = 16y_t + 8$$

and find the particular solution if $y_0 = 1$.

Solution: Divide both sides of the equation $8y_{t+1} = 16y_t + 8$ by 8 to obtain

$$y_{t+1} = 2y_t + 1$$

It follows from (12.7) that

$$y_t = 2^t\left(y_0 - \frac{1}{1-2}\right) + \left(\frac{1}{1-2}\right)$$

Simplifying the expression above we get

$$y_t = 2^t(y_0 + 1) - 1$$

If $y_0 = 1$, then

$$\begin{aligned} & y_t = 2^t(1+1) - 1 = 2^{t+1} - 1 \\ \text{If } t = 1, \quad & y_t = 2^{1+1} - 1 = 3 \\ \text{If } t = 2, \quad & y_t = 7 \\ \text{If } t = 3, \quad & y_t = 15 \end{aligned}$$

The behavior of the solution sequence is given in Figure 12.1. The solution sequence in this example diverges to $+\infty$:

$$y_0 = 1,\ y_1 = 3,\ y_2 = 7,\ y_3 = 15 \text{ and so forth}$$

EXAMPLE 12.2 Solve the first-order difference equations below, calculate the first five values of the solution sequence, and determine the behavior of the solution sequence.

$$8y_{t+1} = 4y_t - 2, \quad y_0 = 3.5 \tag{12.8}$$

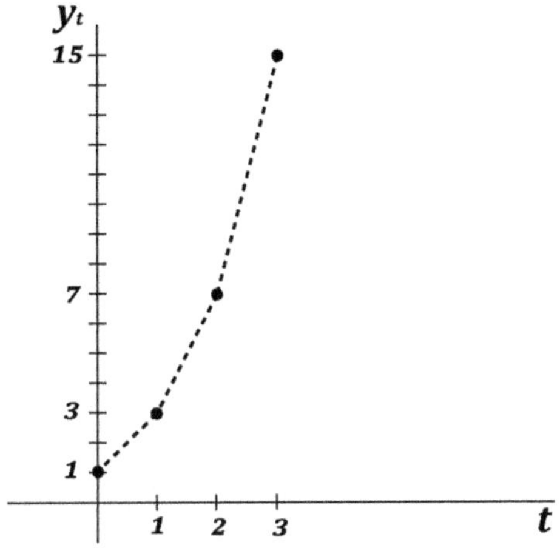

Figure 12.1

Solution: Divide each side of equation (12.8) by 8

$$y_{t+1} = 0.5y_t - 0.25$$

It follows from (12.7) that

$$y_t = 0.5^t \left(y_0 - \frac{-0.25}{1-0.5} \right) + \left(\frac{-0.25}{1-0.5} \right)$$

or

$$y_t = 0.5^t (y_0 + 0.5) - 0.5$$

If $y_0 = 3.5$, then

$$y_t = 0.5^t \cdot 4 - 0.5$$

Thus, we have

$$y_0 = 3.5$$
$$y_1 = 1.5$$
$$y_2 = 0.5$$
$$y_3 = 0$$
$$y_4 = -0.25$$
$$y_5 = -0.375$$
$$y_6 = -0.4375$$

The solution sequence is shown in Figure 12.2.

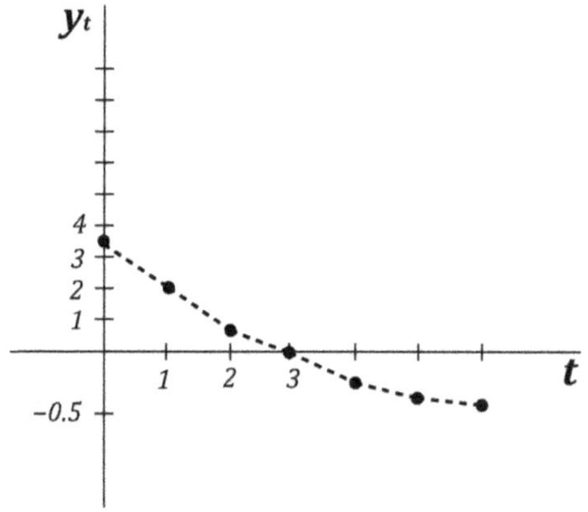

Figure 12.2

Table 12.1 Behavior of the solution sequence $y_{t+1} = ay_t + c$

Case	a ($a \neq 0$)	c	y_0	*Behavior of solution sequence*
(1)	$a = 1$	$c = 0$		constant: $y_t = y_0$
(2)	$a = 1$	$c \neq 0$		diverges to $+\infty$ or $-\infty$
(3)	$a \neq 1$		$y_0 = y^*$	constant: $y_t = y^*$
(4)	$a \leq -1$		$y_0 \neq y^*$	diverges, oscillates
(5)	$-1 < a < 1$		$y_0 \neq y^*$	converges to y^*
(6)	$a > 1$			diverges to $+\infty$ or $-\infty$

As you can see from Figure 12.2, the solution sequence in Example 12.2 converges monotonically down to $y^* = -0.5$. The first-order difference equation in question is said to be globally asymptotically stable whenever the solution converges to the equilibrium state as $t \to \infty$. Table 12.1 illustrates the behavior of the solution sequence for different values of a, c and y_0. The six cases considered in Table 12.1 are reflected in Figure 12.3.

We can sum up the results that we have obtained so far in the theorem below.

THEOREM 12.1: *The linear first-order difference equation with constant coefficients*

$$y_{t+1} = ay_t + c \quad \text{for } a \neq 0,\ t = 0, 1, 2, \ldots$$

has the solution

$$y_t = \begin{cases} a^t (y_0 - y^*) + y^*, & a \neq 1 \\ y_0 + ct, & a = 1 \end{cases}$$

with $y^* = \dfrac{c}{1-a}$. *The solution remains constant if* $y_0 = y^*$. *The sequence converges to* y^* *if* $|a| < 1$. *Otherwise, the solution sequence diverges.*

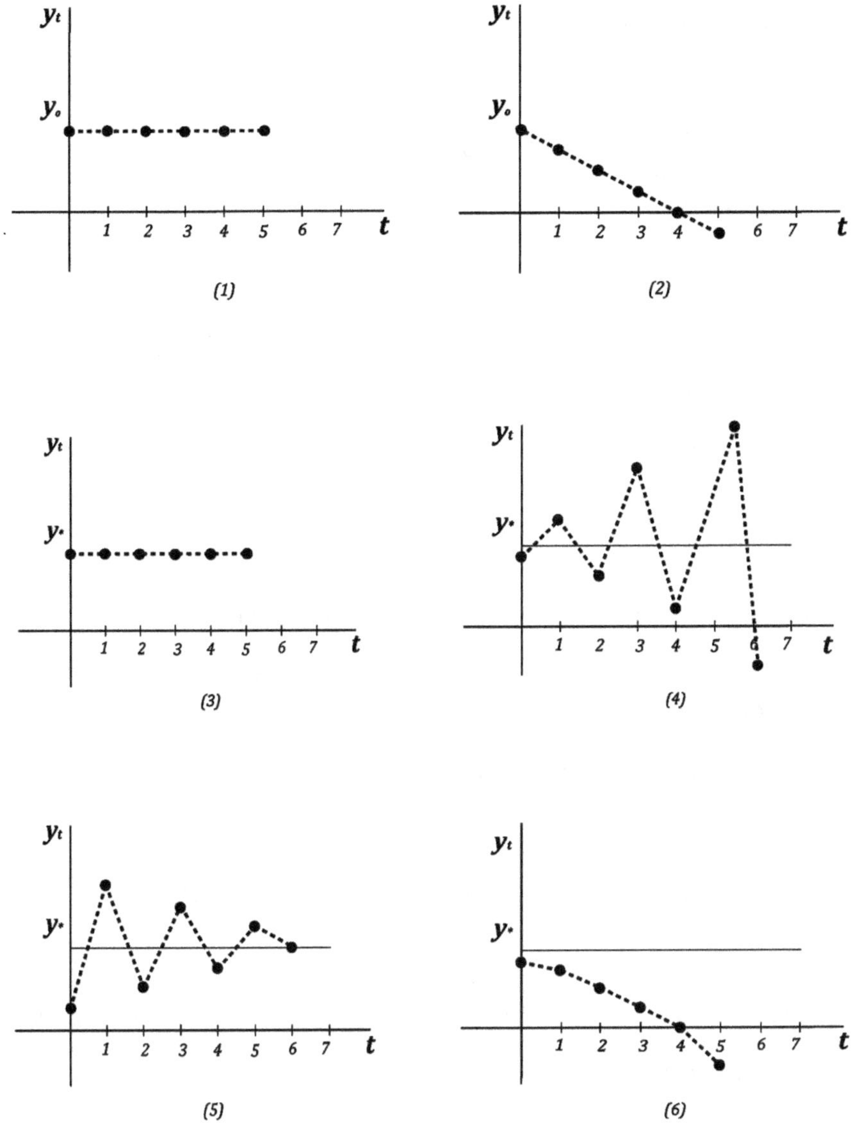

Figure 12.3

12.1.2 Linear second-order difference equations

In this section we will discuss methods of solving linear second-order difference equations with constant coefficients.

A second-order difference equation with constant coefficients can be written as follows:

$$y_{t+2} + a_1 y_{t+1} + a_2 y_t = c(t) \tag{12.9}$$

Now, assuming that $c(t) = 0$ we have

$$y_{t+2} + a_1 y_{t+1} + a_2 y_t = 0 \tag{12.10}$$

Equation (12.10) is referred to as a general homogeneous linear second-order difference equation with constant coefficients.

In order to solve the equation

$$y_{t+2} + a_1 y_{t+1} + a_2 y_t = 0$$

we form the so-called auxiliary equation

$$m^2 + a_1 m + a_2 = 0$$

and find its roots m_1 and m_2:

$$m_{1,2} = \frac{-a_1 \pm \sqrt{a_1^2 - 4a_2}}{2}$$

There are three distinct cases, which are reflected in the theorem below.

THEOREM 12.2: *The general solution of the homogeneous second-order difference equation $y_{t+2} + a_1 y_{t+1} + a_2 y_t = 0$ is as follows:*

(a) *If the auxiliary equation has two distinct real roots, then*

$$y_t = c_1 m_1^t + c_2 m_2^t$$

(b) *If the auxiliary equation has one real double root, $m_1 = m_2 = m$, then*

$$y_t = (c_1 + c_2 t) m^t$$

(c) *If the auxiliary equation has two complex roots $m_1 = a + bi$ and $m_2 = a - bi$, then*

$$y_t = r^t [c_1 \cos(\theta t) + c_2 \sin(\theta t)]$$

where

$$r = \sqrt{a^2 + b^2} \quad \textit{and} \quad \tan\theta = \frac{a}{b}$$

Note that, if the initial conditions (y_0 and y_1) are specified, then the constants c_1 and c_2 can be uniquely determined in all cases.

EXAMPLE 12.3 Obtain the general solution for the following second-order difference equations:

(a) $y_{t+2} - 2y_{t+1} + y_t = 0$
(b) $y_{t+2} + 5y_{t+1} + 6y_t = 0$

Solution:

(a) The auxiliary equation is

$$m^2 - 2m + 1 = (m-1)^2 = 0$$

so $m = 1$ is a double real root. The general solution is

$$y_t = (c_1 + c_2 t) \cdot 1^t = c_1 + c_2 t$$

(b) The auxiliary equation is

$$m^2 + 5m + 6 = (m+2)(m+3) = 0$$

The roots of the auxiliary equation are $m_1 = -2$ and $m_2 = -3$. Thus, the general solution is

$$y_t = c_1(-2)^t + c_2(-3)^t$$

12.1.3 The cobweb model

Economists often assume that supply adjusts to demand instantaneously. In reality this is not always the case. For instance, in agricultural markets there is a lag between the time a crop is planted and when it goes to the market. Farmers plant based on the price of the commodity when it goes into the ground and sell based on the price when it is harvested. The cobweb model allows us to give a quantitative representation of the delayed response and resulting oscillations.

Suppose that the supply and demand is governed by the following two equations:

$$Q_t^{\mathrm{d}} = a - bP_t$$
$$Q_t^{\mathrm{s}} = -c + dP_{t-1}$$

where $a,\ b,\ c,\ d > 0$.

Note that the supply depends on the lagged price. Assuming that the market clears at each time period, we have

$$Q_t^{\mathrm{d}} = Q_t^{\mathrm{s}} = Q_t$$

The linear difference equation for P_t,

$$bP_t = a + c - dP_{t-1}$$

yields the equilibrium value of

$$P^* = \frac{a+c}{b+d}$$

Similarly, the difference equation for Q_t,

$$bQ_t = -cb + ad - dQ_{t-1}$$

produces the equilibrium value of

$$Q^* = \frac{ad - bc}{b + d}$$

The resulting difference equations can be solved using the techniques presented above. Let us look at the qualitative behavior of the solution. The equilibrium point is

$$(P^*, Q^*) = \left(\frac{a + c}{b + d}, \frac{ad - bc}{b + d}\right)$$

The multiplier in our equations is $-\frac{d}{b}$. Therefore the multiplier in our linear difference equations is negative. The resulting solution is oscillatory. Depending on the absolute value of the multiplier, the solutions either converge to the equilibrium (when $d < b$) or diverge (when $d > b$). The equilibria are stable in the first case and unstable in the second case.

PROBLEMS FOR SECTION 12.1

1. Find the general solution for the difference equations below

   (a) $y_{t+2} - 8y_{t+1} + 16y_t = 0$
   (b) $y_{t+2} - 4y_{t+1} + 4y_t = 0$
   (c) $y_{t+2} + 8y_{t+1} + 16y_t = 0$
   (d) $y_{t+2} + 4y_{t+1} + 4y_t = 0$
   (e) $y_{t+2} + 5y_{t+1} + 6y_t = 0$
   (f) $y_{t+2} + 2y_{t+1} - 3y_t = 0$
   (g) $y_{t+2} - 6y_{t+1} + 5y_t = 0$
   (h) $y_{t+2} + y_{t+1} + y_t = 0$

2. Obtain the general solution for the difference equation

$$y_{t+2} - 5y_{t+1} + 6y_t = 0$$

   and the particular solution if $y_0 = 0$ and $y_1 = 1$.

12.2 Differential equations

12.2.1 Classification of differential equations

A differential equation is an equation for a function of one or several variables that contains derivatives, either ordinary derivatives or partial derivatives.

A differential equation that involves derivatives of a function of one variable is referred to as an ordinary differential equation. For example, equation (12.11) is an ordinary differential equation (ODE):

$$\frac{dy}{dx} = 3x^2 + 2x + 1 \tag{12.11}$$

A differential equation that involves derivatives of a function of several variables is known as a partial differential equation (PDE). For instance, equation (12.12) is a partial differential equation (PDE):

$$\frac{\partial y}{\partial x}+\left(\frac{\partial^3 y}{\partial w^3}\right)^2+2x+3w=0 \tag{12.12}$$

The order of a differential equation is the order of the highest derivative that appears in the differential equation. Note that equation (12.11) is an ODE of order one and equation (12.12) is a PDE of order three. The degree of a differential equation is the power of the highest derivative term. Table 12.2 provides several examples of differential equations.

Some differential equations, say equation (12.11), can be written in the form

$$\frac{d^n y}{dx^n}=f(x) \tag{12.13}$$

If a differential equation can be written in the form above, its solution is derived by integration or successive integration.

EXAMPLE 12.4 Find the general solution of the differential equation

$$\frac{dy}{dx}=3x^2+2x+1$$

Solution: If

$$\frac{dy}{dx}=3x^2+2x+1$$

Table 12.2 Examples of differential equations

Equation	*Type*	*Order*	*Degree*
$\frac{dy}{dx}=2x^2+3$	ODE	1	1
$\frac{d^2y}{dx^2}+\left(\frac{dy}{dx}\right)^5+2y=0$	ODE	2	1
$\left(\frac{d^2y}{dx^2}\right)^5+\frac{dy}{dx}+2y=0$	ODE	2	5
$\frac{d^3y}{dx^3}+\left(\frac{d^2y}{dx^2}\right)^3+2xy=0$	ODE	3	1
$\frac{\partial z}{\partial x}+\left(\frac{\partial^3 y}{\partial w^3}\right)^2+2x+3w=0$	PDE	3	2
$\left(\frac{\partial^3 z}{\partial x^3}\right)^5+\left(\frac{\partial z}{\partial x}\right)^7+xz=0$	PDE	3	5

then

$$y = \frac{3x^3}{3} + \frac{2x^2}{2} + x + c = x^3 + x^2 + x + c.$$

Example 12.5 Find the general solution of the differential equation

$$\frac{d^2y}{dx^2} = x + 1$$

Solution: If $\frac{d^2y}{dx^2} = x + 1$, then

$$\frac{dy}{dx} = \frac{x^2}{2} + x + c_1$$

and

$$y = \frac{1}{6}x^3 + \frac{1}{2}x^2 + c_2 + c_1 x$$

There are quite a few types of ODEs whose solutions cannot be found by straightforward integration. We will consider methods for solving the following two types of differential equations: separable differential equations and homogeneous differential equations.

12.2.2 Separable differential equations

If a differential equation can be written in the form

$$\frac{dy}{dx} = f(x)g(y) \tag{12.14}$$

where $f(x)$ is a function of x only and $f(y)$ is a function of y only, then we call the differential equation *separable*. The general solution of the equation in question can be obtained by means of integration:

$$\int \left(\frac{1}{g(y)} \right) dy = \int f(x)\, dx \tag{12.15}$$

Alternatively, if a differential equation can be written in the form

$$h(y)\, dy = k(x)\, dx \tag{12.16}$$

then the equation in question is referred to as *separable* and can be solved by means of integration.

Example 12.6 Solve the equation

$$\frac{dy}{dx} = \frac{x^2}{y^2} \tag{12.17}$$

Solution: It follows from equation (12.17) that

$$dy = \frac{x^2}{y^2}\, dx$$

or

$$y^2\, dy = x^2\, dx$$

The variables are separated and integrating gives us

$$\int y^2\, dy - \int x^2\, dx = c$$
$$\frac{y^3}{3} - \frac{x^3}{3} = c$$
$$y^3 - x^3 = c$$

Note that the constant in the last equation is $3c$ in terms of the preceding equation. However, because c is an arbitrary constant it is easier to introduce a new constant, say $c = a/3$ and then rename a into c. It is more convenient to deal with a simpler expression. This sort of renaming often happens when we deal with the arbitrary constants.

Example 12.7 Solve the equation

$$(2 - y)\, dy + (2 - x)\, dx = 0 \tag{12.18}$$

Solution: The variables in equation (12.18) are already separated. Integrating the equation we get

$$\int (2 - y)\, dy + \int (2 - x)\, dx = c$$
$$2y - \frac{y^2}{2} + 2x - \frac{x^2}{2} = c$$
$$4y - y^2 + 4x - x^2 = c$$

Example 12.8 Suppose the interest rate is $100 \cdot r$ percent compounded continuously and $P(t)$ is the amount of money in the bank at time t. Then

$$\frac{dP}{dt} = r \cdot P$$

Note that the equation above is a separable differential equation. Separating the variables we get

$$\frac{dP}{P} = r \cdot dt$$

And, integrating,

$$\int \frac{dP}{P} = \int r\,dt$$

$$\ln P = rt + c$$

Therefore, we must have

$$e^{\ln P} = e^{rt+c}$$

or

$$P = e^{rt+c} = e^{rt} \cdot e^{c} = k\,e^{rt}$$

Now, let us say that the amount of money in the bank at time $t = 0$ is P_0. Then

$$P_0 = k\,e^0 = k$$

and

$$P = P_0\,e^{rt}$$

Note that here we have derived the formula above that is already very familiar to you through a differential equation.

12.2.3 Homogeneous differential equations

A differential equation of the form

$$f(x, y)\,dx + g(x, y)\,dy = 0$$

is said to be homogeneous if f and g are homogeneous functions of the same degree in both x and y (i.e. $f(kx, ky) = k^n f(x, y)$ and $g(kx, ky) = k^n g(x, y)$, where $k = \text{const}$). If a differential equation is homogeneous, we can use the substitution below

$$y = wx \quad \text{or} \quad x = u\,y \quad \text{where} \quad w = w(x, y) \quad \text{and} \quad u = u(x, y)$$

to separate the variables.

Recall that if $y = wx$ and $x = u\,y$ then

$$dy = w\,dx + x\,dw$$

and

$$dx = u\,dy + y\,du$$

At first, it is not easy to see how these substitutions help us in solving the equation. We will show the process for the y substitution (x substitution works similarly). From the original equation we find

$$\begin{aligned} f(x,y)\,dx+g(x,y)\,dy &= f(x,wx)\,dx+g(x,wx)(w\,dx+x\,dw) \\ &= x^n[f(1,w)\,dx+g(1,w)(w\,dx+x\,dw)] \\ &= 0 \end{aligned}$$

Now cancelling the x^n factor we find

$$f(1,w)\,dx+g(1,w)(w\,dx+x\,dw)=0$$

from which we arrive at the following without too much difficulty:

$$x\frac{dw}{dx}=-w-\frac{f(1,w)}{g(1,w)}$$

The equation above is a separable one, which now can be solved by using the method from previous subsections.

12.2.4 Linear first-order differential equations

Linear differential equations play a very important role in many areas of applied mathematics, physics, engineering and biology. The same is true for economic disciplines. In this section we consider only first-order linear ODEs, as already classifying the behavior of second-order ODEs presents a significant challenge.

Linear first-order ODEs have the following form:

$$y'+p(x)y=q(x)$$

We solve the equation above using the so-called integrating factor method. The idea of the method is to multiply both sides of an equation by the same function $\mu(x)$ such that the resulting equation

$$\mu(x)y'+\mu(x)p(x)y=q(x)\mu(x)$$

has a complete derivative on the right:

$$\mu(x)y'+\mu(x)p(x)y=(\mu(x)y)'$$

Comparing the left-hand and right-hand sides of the above equation we find

$$\mu'=\mu\cdot p$$

or

$$(\ln\mu)'=p$$

which gives us

$$\mu(x) = \exp\left(\int p(x)\,dx\right)$$

It is worth noting that because of the way we use function μ – namely we multiply both sides of an equation by it – the choice of the constant in the indefinite integral above is irrelevant. Now our linear equation is reduced to

$$(\mu y)' = \mu q$$

Usual integration gives us

$$y(x) = \frac{1}{\mu(x)} \int \mu(x) q(x)\,dx$$

Here is an example of a linear differential equation.

EXAMPLE 12.9 Solve

$$y' = -4xy + x$$

This can be rearranged to

$$y' + 4xy = x$$

As in the general case above we will look for an appropriate integrating factor. For this case it is

$$e^{\int 4x\,dx} = e^{2x^2}$$

as multiplying both sides by this integrating factor gives us

$$e^{2x^2} y' + 4x\,e^{2x^2} y = e^{2x^2} x$$

with the left-hand side being a complete derivative:

$$e^{2x^2} y' + 4x\,e^{2x^2} y = (e^{2x^2} y)'$$

Applying integration to both sides yields

$$e^{2x^2} y = \int e^{2x^2} x\,dx = \frac{1}{4}\,e^{2x^2} + c$$

The general solution is

$$y = \frac{1}{4} + c\,e^{-2x^2}$$

EXAMPLE 12.10 Solve

$$y' = \frac{y}{x} + 5$$

Collecting the linear terms with y on the left we have

$$y' - \frac{y}{x} = 5$$

The integrating factor is

$$e^{-\int \frac{1}{x}\, dx} = e^{-\ln x} = \frac{1}{x}$$

Multiplying both sides by it we obtain

$$\frac{1}{x} y' - \frac{y}{x^2} = \frac{5}{x}$$

or

$$\left(\frac{y}{x}\right)' = \frac{5}{x}$$

Integrating the above equation gives us

$$\frac{y}{x} = \int \frac{5}{x}\, dx = 5 \ln x + c$$

Solving for y, we find

$$y = 5x \ln x + cx$$

An important special case of linear equations is the one with constant coefficients.

EXAMPLE 12.11 Solve

$$y' - 3y = -6$$

with the initial condition $y(0) = 1$.

Solution: The integrating factor for linear equations with constant coefficients is always an exponential function. Let us try finding the appropriate integrating factor of the form e^{ax}. Multiplying both sides by it, we find

$$y'\, e^{ax} - 3\, e^{ax}\, y = -6\, e^{ax}$$

It is not difficult to check that the left-hand side is a complete derivative if $a = -3$. Then our equation becomes

$$(y\, e^{-3x})' = -6\, e^{-3x}$$

The integration results in

$$y\,e^{-3x} = -\int 6\,e^{-3x}\,dx = 2\,e^{-3x} + c$$

Equivalently,

$$y = 2 + c\,e^{3x}$$

Substituting $x = 0$ results in

$$y(0) = 2 + c = 1$$

according to the initial condition. That allows us to find c, $c = -1$. The solution to the initial value problem is

$$y(x) = 2 - e^{3x}$$

PROBLEMS FOR SECTION 12.2

1. Solve the following differential equations
   (a) $y' = 3x - x^2$
   (b) $y' = 2\,e^x + 3x - \dfrac{1}{x}$
   (c) $y'' = 2x - 6x^4$
   (d) $y'' = 3\,e^x - 2x + 1$
   (e) $y' = x^2 \tan(y)$
   (f) $y' = (2x - 1)\sqrt{y}$
   (g) $(x + y)\,dx + (x - y)\,dy = 0$
   (h) $xy' + y = 4x^3$
   (i) $y' + 4y = 3$
   (j) $y' = 1 + 2y$
2. Solve the following differential equations
   (a) $y' = 4x - 3x^2$
   (b) $y' = \sin(x) - x^2 + \dfrac{3}{x}$
   (c) $y'' = 4 + 3x^2 - 5x^3$
   (d) $y'' = 2\,e^x - 2x + 3$
   (e) $y' = (2x - x^3)(y^3 - y)$
   (f) $y' = (x^3 + 2x^2 - 1)\,e^y$
   (g) $(x^2 + y^2)\,dx + (x^2 - y^2)\,dy = 0$
   (h) $xy' - y = 2x$

(i) $y' - 3y = 2$

(j) $y' = 4 - 3y$

3. Solve the initial value problems below.

(a) $y' = 2x - x^2$, $y(0) = 0$

(b) $y' = (x^2 - x)y^2$, $y(0) = 1$

(c) $y' = xy - 3$, $y(1) = 2$

(d) $y' - 3y = 2$, $y(0) = 4$

4. Solve the initial value problems below.

(a) $y' = 1 + 4x - x^3$, $y(0) = 0$

(b) $(x + y)\,dx + (x - y)\,dy' = 0$, $y(1) = 1$

(c) $y' = xy$, $y(1) = 2$

(d) $y' - 2y = 1$, $y(0) = 3$

Answers to odd-numbered problems

CHAPTER 1

Section 1.1

1. (a) $S = \{x \mid x \in \mathbb{R}\}$
   (b) $S = \{x \mid x \in \mathbb{N}\}$
   (c) $S = \{x \mid x \in \mathbb{Z}^+\}$
   (d) $S = \{x \in \mathbb{R} \mid x > 76\}$
   (e) $S = \{x \in \mathbb{N} \mid x < 22\}$
   (f) $S = \varnothing$

3. (a) $P(B_1) = \{\varnothing, \{a\}\}$
   (b) $P(B_2) = \{\varnothing, \{a\}, \{b\}, \{a, b\}\}$
   (c) $P(B_3) = \{\varnothing, \{1\}, \{2\}, \{3\}, \{1, 2\}, \{1, 3\}, \{2, 3\}, \{1, 2, 3\}\}$
   (d) $P(B_4) = \{\varnothing, \{5\}, \{7\}, \{9\}, \{5, 7\}, \{5, 9\}, \{7, 9\}, \{5, 7, 9\}\}$
   (e) $P(B_5) = \{\varnothing, \{a, b\}, \{a, c\}, \{\{a, b\}, \{a, c\}\}\}$

5. To prove that $A = B$ we need to show that: (a) if x is in A then x is in B and (b) if x is in B then x is in A. To prove (a) recall that by assumption $A \subset B$. Hence, every element x of A is also an element of B. Statement (b) can be proved in a similar fashion.

Section 1.2

1. (a) $x_{1,2} = -2, -5$
   (b) $x_{1,2} = -3, -5$
   (c) $x_{1,2} = \frac{3}{2}(3 \pm \sqrt{5})$
   (d) $x_{1,2} = -2, 8$
   (e) $x_{1,2} = -3, 5$
   (f) $x_{1,2} = -\frac{3}{2}, 10$
   (g) $x_{1,2} = \frac{3}{2}$

CHAPTER 2

Section 2.2

1. (a) 8 (b) 3 (c) 2 (d) 8 (e) 4 (f) 10 (g) 2 (h) 2

Section 2.4

1. (a) $\sum_{i=1}^{100} i$ (b) $\sum_{i=1}^{100} \frac{1}{i}$ (c) $\sum_{n=2}^{10} \frac{1}{n^2}$ (d) $\sum_{n=1}^{99} n(n+1)$

   (e) $\sum_{j=1}^{11} j^2$ (f) $\sum_{n=1}^{18} \frac{w^{2n-1}}{2n-1}$ (g) $\sum_{i=1}^{99} (-1)^{i+1} i$

   (h) $\sum_{n=1}^{12} n(n+3)$ (i) $\sum_{n=1}^{96} \frac{n(n+1)}{(n+2)(n+3)}$

3. (a) True
   (b) True
   (c) True
   (d) False
   (e) False

5. We observe that for every term $i+j$ on the left-hand side there is exactly one corresponding term $j+i=i+j$ on the right-hand side with the same value $i+j$ and vice versa. That proves the result. The summation is over the same set of pairs (i, j), they are just listed in a different order.

CHAPTER 3

Sections 3.3

1. (a) $4 \ln 5$
   (b) $\frac{2}{3} \ln 5$
   (c) $-3 \ln 5$
   (d) $-\frac{1}{2} \ln 5$

3. (a) True
   (b) True
   (c) True
   (d) False
   (e) False
   (f) False

5. In $\frac{\ln(2)}{\ln(1.07)} = 10.2$ years

CHAPTER 4

Section 4.3

1. Find the first derivative of the following functions:

   (a) $y'(x) = 20x^3 - 12x^2 + 6x - 21$

   (b) $g'(x) = 5\,e^x + \frac{1}{x} - x^{-0.5}$

(c) $h'(x) = -\dfrac{bc}{(cx)^2}$

(d) $k'(x) = a[3x^2(x^2+1)(x+1) + 2x(x^3+1)(x+1) + (x^3+1)(x^2+1)]$

(e) $w'(x) = 4ax^3 + b[(cx^2+2) + 2cx(x+1)]$

(f) $u'(x) = \dfrac{-2x^3 - 12x^2 + 2}{(x^3+2)^2}$

(g) $f'(x) = e^x(\ln x + 2)(x^4+3) + \dfrac{1}{x}(e^x+1)(x^4+3) + 4x^3(e^x+1)(\ln x + 2)$

(h) $r'(x) = \dfrac{\frac{1}{x}(e^x+2) - e^x(\ln x + 1)}{(e^x+2)^2}$

(i) $$\begin{aligned} t'(x) = {} & 2x(x^3-4)(x^4+5)(x^5-6)(x^6+7) \\ & + 3x^2(x^2+3)(x^4+5)(x^5-6)(x^6+7) \\ & + 4x^3(x^2+3)(x^3-4)(x^5-6)(x^6+7) \\ & + 5x^4(x^2+3)(x^3-4)(x^4+5)(x^6+7) \\ & + 6x^5(x^2+3)(x^3-4)(x^4+5)(x^5-6) \end{aligned}$$

3. (a) $\dfrac{dy}{dx} = -160\dfrac{(4x^3+4)}{(x^4+4x+5)^9}$

(b) $\dfrac{dy}{dx} = 99(5x^4+2x)(x^5+x^2)^{98}$

(c) $\dfrac{dy}{dx} = \dfrac{1}{3}\left(\dfrac{x-1}{x+6}\right)^{-\frac{2}{3}} \times \dfrac{7}{(x+6)^2}$

(d) $\dfrac{dy}{dx} = \dfrac{5x^4}{2\sqrt{x^5+17}}$

(e) $\dfrac{dy}{dx} = 75(1+x^{24})(25x+x^25)^2$

(f) $\dfrac{dy}{dx} = 4\left(\dfrac{x+2}{100x^4+2x+10}\right)^3 \dfrac{-800x^3+96}{(100x^4+2x+10)^2}$

(g) $\dfrac{dy}{dx} = -\dfrac{3}{2}\sqrt{\dfrac{x-1}{x+2}}\dfrac{1}{(x-1)^2}$

(h) $\dfrac{dy}{dx} = 6(x^5+x^4+x^3+x^2+x+1)^5(5x^4+4x^3+3x^2+2x+1)$

(i) $\dfrac{dy}{dx} = 3(x^4+2x^2+12)^{-\frac{1}{4}}(x^3+x)$

(j) $\dfrac{dy}{dx} = \dfrac{704x^3}{3}(2x^4+5)^{87}$

5. $\dfrac{dW}{dt} = \dfrac{dh}{dk}\dfrac{dk}{dt}$

7. $\dfrac{d(\mathrm{TC})}{dQ} = \dfrac{9}{4}(525-3Q)^{-\frac{3}{4}}$

Section 4.5

1. (a) $\mathrm{MC}(Q) = Q^2 - 14Q + 64$

   (b) $\mathrm{AC}(Q) = \frac{1}{3}Q^2 - 7Q + 64 + \frac{300}{Q}$

   (c) $Q = 7$

3. $\frac{dY}{Y\,dt} = \frac{b}{a+bt}$, $\frac{dP}{P\,dt} = r$ and $\frac{b}{a+bt} - r$

5. (a) $\frac{dy}{y\,dt} = \frac{4}{t}$

   (b) $\frac{y_{t+1} - y_t}{y_t} = \frac{4t^3 + 6t^2 + 4t + 1}{t^4}$

CHAPTER 5

Section 5.1

1. (a) $x = 3$ is a local minimum
   (b) $x = 6$ is a local minimum
   (c) $x = -5$ is a local minimum
   (d) $x = 0.25$ is a local minimum
   (e) $x = -1$ is a local maximum

3. $Q = 2$; Price $= \$10$

5. To maximize his profits the monopolist should produce at $Q = 2$

$$\frac{d\pi}{dQ} = b(-\alpha)Q^{-\alpha-1} - a$$

$$\frac{d^2\pi}{dQ^2} = b(-\alpha)(-\alpha - 1)Q^{-\alpha-2}$$

$$Q^* = \left(\frac{a}{-b\alpha}\right)^{\frac{1}{-\alpha-1}}$$

7. (a) To maximize profits the perfectly competitive firm Digital Solutions Company should produce at $Q^* = 3$.

Section 5.2

1. $f(x) \approx 1 - 0.5x$

   $\sqrt{0.98} \approx 0.99$ and $\sqrt{0.95} \approx 0.975$

3. $f(x) \approx 3 + \frac{0.5}{3}(x - 1)$

   $\sqrt{8.99} \approx 2.998$ and $\sqrt{9.02} \approx 3.003$

5. $f(x) \approx 1 + \frac{1}{3}x$

   $\sqrt[3]{1.01} \approx 1.003$ and $\sqrt[3]{0.97} \approx 0.99$

7. $1 - \frac{x}{1!} + \frac{x^2}{2!} - \frac{x^3}{3!} + \frac{x^4}{4!} + \cdots + \cdots + \frac{(-1)^n x^n}{n!}$

CHAPTER 6

Section 6.1

1. (a) $\mathbf{ab} = \begin{bmatrix} 9 & 3 & 6 & 12 \\ 15 & 5 & 10 & 20 \\ 6 & 2 & 4 & 8 \\ 3 & 1 & 2 & 4 \end{bmatrix}$

   (b) $\mathbf{ba} = [22]$

3. (a) $\mathbf{AB} = \begin{bmatrix} 3 & 10 & 12 & 5 \\ -37 & 37 & -5 & 12 \\ -24 & 68 & 36 & 28 \end{bmatrix}$

   $k\mathbf{A} = \begin{bmatrix} 16 & 8 \\ -24 & 40 \\ 32 & 64 \end{bmatrix}$

   $k\mathbf{B} = \begin{bmatrix} 32 & 8 & 40 & 8 \\ -40 & 64 & 16 & 24 \end{bmatrix}$

   The dimension of **B** is (2×4), the dimension of **A** is $(3 \times 2) \Rightarrow$ **BA** is undefined.

   (b) $\mathbf{IA} = \mathbf{AI} = \mathbf{A}$ and $\mathbf{IB} = \mathbf{BI} = \mathbf{B}$. The dimension of matrix **I** in **IB** is (2×2).

5. (a) $(\mathbf{I} - \mathbf{A})(\mathbf{I} - \mathbf{A}) = \mathbf{I}^2 - \mathbf{AI} - \mathbf{IA} + \mathbf{A}^2 = \mathbf{I} - \mathbf{A} - \mathbf{A} + \mathbf{A} = \mathbf{I} - \mathbf{A}$

   (b) $\mathbf{A}^3 = \mathbf{A}^2 \cdot \mathbf{A} = \mathbf{A} \cdot \mathbf{A} = \mathbf{A}$

   $\mathbf{A}^4 = \mathbf{A}^2 \cdot \mathbf{A}^2 = \mathbf{A} \cdot \mathbf{A} = \mathbf{A}$

   $\mathbf{A}^5 = \mathbf{A}^4 \cdot \mathbf{A} = \mathbf{A} \cdot \mathbf{A} = \mathbf{A}$

Section 6.5

1. (a) $|A_1| = 21$ (b) $|A_2| = -2$

   (c) $|A_3| = 0$ (d) $|A_4| = 0$

   (e) $|A_5| = 60$ (f) $|A_6| = 60$

   (g) $|A_7| = 0$ (h) $|A_8| = 41$

3. $\mathbf{A}_1^{-1} = \begin{bmatrix} 0.4 & -0.1 \\ -0.2 & 0.3 \end{bmatrix} \qquad \mathbf{A}_2^{-1} = \begin{bmatrix} 1 & 0 & 0 \\ 0 & 1 & 0 \\ 0 & 0 & 1 \end{bmatrix}$

5. (a) $\mathbf{A}' = \begin{bmatrix} 1 & 0 & 0 \\ 0 & 1 & 0 \\ 0 & 0 & 1 \end{bmatrix} \qquad \mathbf{B}' = \begin{bmatrix} 2 & 1 & 3 \\ 3 & 5 & 7 \end{bmatrix}$

(b) $(\mathbf{ABC})' = \begin{bmatrix} 11 & 16 & 24 \\ 16 & 22 & 34 \end{bmatrix}$ and $\mathbf{C}'\mathbf{B}'\mathbf{A}' = \begin{bmatrix} 11 & 16 & 24 \\ 16 & 22 & 34 \end{bmatrix}$

CHAPTER 7

Section 7.2

1. $\mathbf{A}_1$ is positive definite; $\mathbf{A}_2$ is negative definite; $\mathbf{A}_3$ is negative definite; $\mathbf{A}_4$ is positive definite.

Section 7.5

1. (a) Rank$(\mathbf{A}_1) = 2$. Hence, $\mathbf{A}_1$ is not full rank matrix
   (b) Rank$(\mathbf{A}_2) = 2$. Hence, $\mathbf{A}_2$ is not full rank matrix

3. (a) Rank$(\mathbf{A}_1) = 2$. Hence, $\mathbf{A}_1$ is not full rank matrix
   (b) Rank$(\mathbf{A}_2) = 2$. Hence, $\mathbf{A}_2$ is full rank matrix

Section 7.8

1. (a)

$$\mathbf{A} \otimes \mathbf{B} = \begin{bmatrix} 24 & 48 & 60 & 4 & 8 & 10 \\ -60 & 144 & 12 & -10 & 24 & 2 \\ 24 & 132 & 36 & 4 & 22 & 6 \\ 6 & 12 & 15 & 2 & 4 & 5 \\ -15 & 36 & 3 & -5 & 12 & 1 \\ 6 & 33 & 9 & 2 & 11 & 3 \end{bmatrix}$$

(b)

$$\mathbf{B} \otimes \mathbf{A} = \begin{bmatrix} 24 & 4 & 48 & 8 & 60 & 10 \\ 6 & 2 & 12 & 4 & 15 & 5 \\ -60 & -10 & 144 & 24 & 12 & 2 \\ -15 & -5 & 36 & 12 & 3 & 1 \\ 24 & 4 & 132 & 22 & 36 & 6 \\ 6 & 2 & 33 & 11 & 9 & 3 \end{bmatrix}$$

3.

$$(\mathbf{A}\otimes\mathbf{B})' = \mathbf{A}'\otimes\mathbf{B}' = \begin{bmatrix} 0 & 0 & -2 & -1 & 0 & 0 \\ 0 & 0 & 3 & 2 & 0 & 0 \\ 0 & 0 & -4 & -5 & 0 & 0 \\ 2 & 1 & 0 & 0 & 2 & 1 \\ -3 & -2 & 0 & 0 & -3 & -2 \\ 4 & 5 & 0 & 0 & 4 & 5 \\ 0 & 0 & -2 & -1 & 0 & 0 \\ 0 & 0 & 3 & 2 & 0 & 0 \\ 0 & 0 & -4 & -5 & 0 & 0 \end{bmatrix}$$

5.

$$\mathbf{A}\otimes(\mathbf{B}\otimes\mathbf{C}) = (\mathbf{A}\otimes\mathbf{B})\otimes\mathbf{C}$$

$$= \begin{bmatrix} 0 & 0 & 0 & 0 & 0 & 0 & 4 & 6 & -6 & -9 & 8 & 12 \\ 0 & 0 & 0 & 0 & 0 & 0 & 8 & 14 & -12 & -21 & 16 & 28 \\ 0 & 0 & 0 & 0 & 0 & 0 & 2 & 3 & -4 & -6 & 10 & 15 \\ 0 & 0 & 0 & 0 & 0 & 0 & 4 & 7 & -8 & -14 & 20 & 35 \\ -4 & -6 & 6 & 9 & -8 & -12 & 0 & 0 & 0 & 0 & 0 & 0 \\ -8 & -14 & 12 & 21 & -16 & -28 & 0 & 0 & 0 & 0 & 0 & 0 \\ -2 & -3 & 4 & 6 & -10 & -15 & 0 & 0 & 0 & 0 & 0 & 0 \\ -4 & -7 & 8 & 14 & -20 & -35 & 0 & 0 & 0 & 0 & 0 & 0 \end{bmatrix}$$

CHAPTER 8

Section 8.4

1. $a = 50$ and $b = 25$; $\pi^*(50, 25) = \$3115$
3. $a = 50$, $b = 25$ and $d = 20$; $\pi^*(50, 25, 20) = \$3515$
5. (a)

$$q_1 = \frac{10b}{(a+b+2ab)}$$

$$q_2 = \frac{10a}{(a+b+2ab)}$$

(b)

$$\frac{\partial q_1}{\partial b} = \frac{10a}{(a+b+2ab)^2}$$

$$\frac{\partial q_2}{\partial a} = \frac{10b}{(a+b+2ab)^2}$$

CHAPTER 9

Section 9.1

1. $3x$
3. $\dfrac{x^4}{2}$

5. $-\dfrac{1}{3x^3}$
7. $-\dfrac{1}{2x^4}$
9. $e^x + e^2 x$
11. $\dfrac{32}{47}x^{\frac{47}{32}}$
13. $4x^{0.5} + 3x$
15. $\dfrac{1}{12}e^{3x}(4+3e^x)$
17. 0
19. $ax + bx + \dfrac{x^2}{2}$
21. $\dfrac{y^3}{9} + \dfrac{2y^5}{25}$

Section 9.2

1. $81(1+x^2)^6$
3. $\dfrac{27}{8}(1+x^2)^4$
5. $\dfrac{1}{272}(2+17x)^{16}$
7. $-\dfrac{1}{2x^4}$
9. $e^x(x-1)$
11. $\dfrac{2}{3}(3+x)^{1.5}$
13. $e^x(2-2x+x^2)$
15. $\dfrac{1}{27}e^{3x}(2-6x+9x^2)$
17. $5\,e^{7x}\left(\dfrac{x}{7} - \dfrac{1}{49}\right)$
19. $-\dfrac{x^2}{4} + \dfrac{1}{2}x^2\ln(x)$
21. $(1+15t)y$

Section 9.4

1. (a) $\dfrac{2}{3}$
   (b) $\dfrac{1}{4}$
   (c) 0
   (d) 36

(e) $\frac{8}{3}$

(f) 20

3. $\frac{1}{6}$

CHAPTER 10

Section 10.5

1. $\mathrm{PV} \approx \$376{,}889.48$
3. $\mathrm{PV} \approx \$164{,}169.99$
5. $t \approx 18$ years

   (a) $\mathrm{PV} \approx \$555.02$

   (b) $\mathrm{PV} = \$1063.79$

   (c) $\mathrm{PV} = \$1597.08$

   (d) $\mathrm{PV} = \$2281.88$

Section 10.6

1. $\mathrm{PV_a} = \$1{,}317{,}160$; $\mathrm{PV_b} = \$1{,}009{,}859$. Hence, option (b) is the better one.
3. Payoff $= \$9025.15$

CHAPTER 11

Section 11.3

1. (a) $8-4i$

   (b) $13+18i$

   (c) $12-7i$

   (d) $\frac{1+4i}{3+2i} = \frac{11}{13} + \frac{10}{13}i$

   (e) $\frac{1}{2} - \frac{1}{2}i$

   (f) $i^3 = i \cdot i^2 = -i$

   (g) $\sqrt{-25} = 5\sqrt{1} = 5i$

   (h) $-5+2i$

   (i) $2-19i$

   (j) $2-i$

   (k) $-\frac{5}{17} + \frac{14}{17}$

   (l) $\frac{12}{25} + \frac{9}{25}i$

(m) 1

(n) $\sqrt{36} = 6$

3. (a) $\overline{z_1} = 12 + 15i$ and $|\overline{z_1}| = 3\sqrt{41}$

(b) $\overline{z_2} = 4i$ and $|\overline{z_2}| = 4$

(c) $\overline{z_3} = -1 - 2\sqrt{2}i$ and $|\overline{z_3}| = 3$

5. (a) $\text{Arg}(z_1) = \pi$

(b) $\text{Arg}(z_2) = -\dfrac{\pi}{4}$

(c) $\text{Arg}(z_3) = \dfrac{\pi}{4}$

(d) $\text{Arg}(z_4) = \dfrac{\pi}{2}$

7. (a) -1024

(b) $-512\sqrt{3} + 512i$

(c) $16 + 16\sqrt{3}i$

(d) 16

9. (a) $e^{\frac{i\pi}{2}} = \cos\left(\dfrac{\pi}{2}\right) + i\sin\left(\dfrac{\pi}{2}\right)$

(b) $e^{2\pi i} = \cos(2\pi) + i\sin(2\pi)$

(c) $e^{\frac{i\pi}{3}} = \cos\left(\dfrac{\pi}{3}\right) + i\sin\left(\dfrac{\pi}{3}\right)$

(d) $e^{-i\pi} = \cos(\pi) - i\sin(\pi)$

(e) $e^{i\pi} = \cos(\pi) + i\sin(\pi)$

(f) $e^{-i} = \cos(-1) - i\sin(-1)$

CHAPTER 12

Section 12.1

1. (a) $y_t = (a + bt)4^t$

(b) $y_t = (a + bt)2^t$

(c) $y_t = (a + bt)(-4)^t$

(d) $y_t = (a + bt)(-2)^t$

(e) $y_t = c_1(-2)^t + c_2(-3)^t$

(f) $y_t = c_1(-3)^t + c_2$

(g) $y_t = c_1 5^t + c_2$

(h) $y_t = c_1\cos\left(\dfrac{\pi}{3}t\right) + c_2\sin\left(\dfrac{\pi}{3}t\right)$

Section 12.2

1. (a) $y = \dfrac{3}{2}x^2 - \dfrac{x^3}{3} + c$

(b) $y = 2e^x + \frac{3}{2}x^2 - \ln(x) + c$

(c) $y = \frac{x^3}{3} - \frac{x^6}{5} + c_1 x + c_2$

(d) $y = 3e^x - \frac{x^3}{3} + \frac{x^2}{2} + c_1 x + c_2$

(e) $\ln(\sin(y)) = \frac{x^3}{3} + c$

(f) $y = \left[\frac{1}{2}(x^2 - x + c)\right]^2$

(g) $1 - \frac{y^2}{x^2} + 2\frac{y}{x} = \frac{k}{x2}$

(h) $y = x^3 + \frac{c}{x}$

(i) $y = \frac{3}{4} + c\,e^{-4x}$

(j) $y = -\frac{1}{2} + c\,e^{2x}$

3. (a) $y = x^2 - \frac{x^3}{3}$

(b) $\frac{1}{y} = 1 + \frac{x^2}{2} - \frac{x^3}{3}$

(c) $y = -3 + \frac{5}{x}$

(d) $y = -\frac{2}{3} + \frac{14}{3}e^{3x}$

Index

For Product Safety Concerns and Information please contact our EU representative GPSR@taylorandfrancis.com
Taylor & Francis Verlag GmbH, Kaufingerstraße 24, 80331 München, Germany

www.ingramcontent.com/pod-product-compliance
Lightning Source LLC
LaVergne TN
LVHW080809011026
849160LV00015B/1579
9780415776875